第三届艺术与科学
国际作品展作品集

A WORKS COLLECTION OF
THE 3RD ART AND SCIENCE
WORKS INTERNATIONAL EXHIBITION

2012.11.1—2012.11.30

第三届艺术与科学国际作品展
暨学术研讨会筹备办公室　编

中国建筑工业出版社

| Information | Ecology | Wisdom |
| 信息 | 生态 | 智慧 |

ART AND SCIENCE

支持单位 — Support Units

中华人民共和国文化部 — Ministry of Culture of the People's Republic of China
中国科学技术协会 — China Association for Science and Technology
中国文学艺术界联合会 — China Federation of Literary and Art Circles

主办单位 — Organizers

清华大学 — Tsinghua University
中国科学技术馆 — China Science & Technology Museum

承办单位 — Co-organizers

清华大学艺术与科学研究中心 — Art & Science Research Center of Tsinghua University
清华大学美术学院 — Academy of Arts & Design of Tsinghua University
如意科技集团 — Ruyi Tech Group

协办单位 — Cooperation with

九牧集团 — JOMOO Group Co. Ltd.
深圳市福田区政府 — Futian District Government of Shenzhen
三林·生活家集团 — Samling Elegantliving Group

展览时间 — Exhibition Time

2012年11月1日—30日 — November 1, 2012 - November 30, 2012

展览地点 — Exhibition Venue

中国科学技术馆 — China Science & Technology Museum

组织机构

支持单位
中华人民共和国文化部
中国科学技术协会
中国文学艺术界联合会

主办单位
清华大学
中国科学技术馆

承办单位
清华大学艺术与科学研究中心
清华大学美术学院
如意科技集团

协办单位
九牧集团
深圳市福田区政府
三林·生活家集团

合作机构
荷兰 V2 多变媒体协会
法国蓬皮杜艺术中心
奥地利 AEC 电子艺术中心
奥地利林茨大学
美国麻省理工学院媒体实验室
美国罗德岛设计学院
美国帕森斯设计学院
英国皇家艺术学院
德国柏林艺术大学
歌德学院北京分院
荷兰蒙特里安基金会
瑞士文化基金会
中国高等科学技术中心
中央美术学院
中国美术学院
广州美术学院
中国传媒大学

组织委员会

顾问（以姓氏笔画为序）

马国馨	王明旨	白春礼	冯骥才	刘巨德
杨 卫	周其凤	赵忠贤	袁运甫	顾秉林
钱绍武	黄永玉	戚发轫	龚 克	常沙娜
韩美林	靳尚谊			

名誉主席

陈 希　陈吉宁

主席

冯 远　徐延豪　谢维和

执行主席

鲁晓波　殷 皓　邱亚夫

副主席

黄体茂　欧建成　赵 萌　张 敢　苏 丹

秘书长

杨冬江

副秘书长

钱 岩　丁彩玲　任 茜

委员（以姓氏笔画为序）

王旭东	方晓风	任 茜	刘 兵	刘 新
刘振生	李 煜	李砚祖	师丹青	杨 力
肖文陵	邹 欣	张 月	张夫也	张玉银
陈 超	陈 辉	陈 强	陈 楠	陈明辉
林乐成	郑 艺	郑 宁	赵 凯	赵 健
徐迎庆	崔小青	曾成钢	管沄嘉	臧迎春

学术委员会

主席

李政道　冯 远

副主席

王明旨　鲁晓波　黄体茂

委员（以姓氏笔画为序）

马 泉	王希季	王明旨	王胜利	韦尔申
牛志升	方晓风	尹 鸿	卢新华	吕品田
朱幼文	庄惟敏	刘 兵	刘大为	刘元风
刘巨德	许 江	苏 丹	杜大恺	李当岐
李克强	李砚祖	李衍达	杨冬江	杨永善
杨晓阳	吴建平	何 洁	邹建平	汪大伟
张 敢	陈进海	陈积芳	邵大箴	范迪安
杭 间	罗中立	郑 力	郑 宁	郑曙旸
赵 萌	柳冠中	施一公	姜 陆	袁运甫
钱 易	徐勇民	奚静之	诸 迪	常沙娜
崔希栋	康克军	隗京花	鲁晓波	谢维和
靳尚谊	蔡 军	廖 红	潘公凯	

山本圭吾（日本）　　张尕（美国）
埃里克斯·阿瑞安斯（荷兰）　　安东尼·邓恩（英国）
克莱尔·法约尔（法国）　　戴维·波恩（美国）
尅丝·阿姆斯特朗（澳大利亚）　草原真知子（日本）
萨班·海默巴克（德国）

总策展人

鲁晓波

策展顾问

安东尼·邓恩（英国）
克莱尔·法约尔（法国）
草原真知子（日本）
杰斯·豪瑟（法国）
奥瑞·卡茨（以色列和澳大利亚）
皮埃尔·凯勒（瑞士）
萨班·海默巴克（德国）
张尕（美国）

ORGANIZATION

Support Units

Ministry of Culture of the People's Republic of China

China Association for Science and Technology

China Federation of Literary and Art Circles

Organizers

Tsinghua University

China Science & Technology Museum

Co-organizers

Art & Science Research Center of Tsinghua University

Academy of Arts & Design of Tsinghua University

Ruyi Tech Group

Cooperation with

JOMOO Group Co. Ltd.

Futian District Government of Shenzhen

Samling Elegantliving Group

In Cooperation With

V2, Institute for Unstable Media (Netherlands)

Centre Pompidou (France)

Arts Electronica Center (Austria)

Linz University (Austria)

MIT Media Lab (USA)

Rhode Island School of Design (USA)

Parsons School of Design (USA)

Royal Academy of Art (UK)

Universitaet der Künste Berlin (Germany)

Goethe Institut, Peking (Germany)

Mondriaan Foundation (Netherlands)

Switzerland Culture Foundation (Switzerland)

China Center of Advanced Science and Technology (PRC)

China Central Academy of Fine Arts (PRC)

China Academy of Art (PRC)

Guangzhou College of Fine Arts (PRC)

Communication University of China (PRC)

ORGANIZATION COMMITTEE

Advisors

(on the Order of Strokes of Chinese Characters)

Ma Guoxin	Wang Mingzhi	Bai Chunli
Feng Jicai	Liu Jude	Yang Wei
Zhou Qifeng	Zhao Zhongxian	Yuan Yunfu
Gu Binglin	Qian Shaowu	Huang Yongyu
Qi Faren	Gong Ke	Chang Shana
Han Meilin	Jin Shangyi	

Honorary Chairman

Chen Xi   Chen Jining

Chairman

Feng Yuan   Xu Yanhao   Xie Weihe

Executive Chairman

Lu Xiaobo   Yin Hao   Qiu Yafu

Vice Chairman

Huang Timao	Ou Jiancheng	Zhao Meng
Zhang Gan	Su Dan	

General Secretary

Yang Dongjiang

Vice General Secretary

Qian Yan   Ding Cailing   Ren Qian

Committee Menbers

(on the Order of Strokes of Chinese Characters)

Wang Xudong	Fang Xiaofeng	Ren Qian
Liu Bing	Liu Xin	Liu Zhensheng
Li Yu	Li Yanzu	Shi Danqing
Yang Li	Xiao Wenling	Zou Xin
Zhang Yue	Zhang Fuye	Zhang Yuyin
Chen Chao	Chen Hui	Chen Qiang
Chen Nan	Chen Minghui	Lin Lecheng
Zheng Yi	Zheng Ning	Zhao Kai
Zhao Jian	Xu Yingqing	Cui Xiaoqing
Zeng Chenggang	Guan Yunjia	Zang Yingchun

ACADEMIC COMMITTEE

Chairman

Li Zhengdao   Feng Yuan

Vice Chairman

Wang Mingzhi   Lu Xiaobo   Huang Timao

Committee Members

(on the Order of Strokes of Chinese Characters)

Ma Quan	Wang Xiji	Wang Mingzhi
Wang Shengli	Wei Ershen	Niu Zhisheng
Fang Xiaofeng	Yin Hong	Lu Xinhua
Lv Pintian	Zhu Youwen	Zhuang Weimin
Liu Bing	Liu Dawei	Liu Yuanfeng
Liu Jude	Xu Jiang	Su Dan
Du Dakai	Li Dangqi	Li Keqiang
Li Yanzu	Li Yanda	Yang Dongjiang
Yang Yongshan	Yang Xiaoyang	Wu Jianping
He Jie	Zou Jianping	Wang Dawei
Zhang Gan	Chen Jinhai	Chen Jifang
Shao Dazhen	Fan Di'an	Hang Jian
Luo Zhongli	Zheng Li	Zheng Ning
Zheng Shuyang	Zhao Meng	Liu Guanzhong
Shi Yigong	Jiang Lu	Yuan Yunfu
Qian Yi	Xu Yongmin	Xi Jingzhi
Zhu Di	Chang Shana	Cui Xidong
Kang Kejun	Kui Jinghua	Lu Xiaobo
Xie Weihe	Jin Shangyi	Cai Jun
Liao Hong	Pan Gongkai	

Keigo Yamamoto (Japan)

Ga Zhang (USA)   Alex Adriaansens (Netherland)

Anthony Dunne (UK)   Claire Fayolle (France)

David Bowen (USA)   Keith Armstrong (Australia)

Machiko Kusahara (Japan)

Sabine Himmelsbach (Germany)

General Planner

Lu Xiaobo

Curatorial Consultant

Anthony Dunne(UK)

Claire Fayolle (France)

Machiko Kusahara(Japan)

Jens Hauser (France)

Oran Catts (Israel and Australia)

Pierre Keller (Switzerland)

Sabine Himmelsbach (Germany)

Ga Zhang (USA)

致辞

2001年，在清华大学九十周年校庆之际，李政道先生和吴冠中先生发起举办了首届"艺术与科学国际作品展暨学术研讨会"，充分彰显了艺术之美与科学之美的碰撞与融合。艺术与科学，这一深邃而开放的命题，吸引着越来越多的学者潜心地探索、研究和实践。

艺术与科学，虽然角度不同、方法有异，但其目标都是对真理的求索。我们常说科学求真、人文求善、艺术求美，其实艺术中也有科学性，科学中亦蕴涵着艺术性，两者都是推动人类文明发展和社会进步的重要力量。

科学与人文、艺术的相互补给、相互映照，也为人才培养事业提供了源源不断的动力。作为一所综合性、研究型、开放式的大学，清华大学不断地推动着人文精神与自然科学的融合，促进艺术学科与理工学科的互动，提倡开展交叉性创新研究，鼓励学生在接受专业教育的同时，加强其他学科知识的贯通融会，提升综合能力、塑造完整人格。

在清华大学新百年的开端，2012年的初冬，"第三届艺术与科学国际作品展暨学术研讨会"隆重开幕。这必将为促进未来高等教育领域中艺术与科学的融合，为人才培养、学术研究以及人类生存质量的提升，开拓更为广阔的空间。衷心希望艺术与科学展能历久弥新，不断焕发新的生命力，为科学与艺术的进步、为中外思想和文化交流贡献一份力量。

陈吉宁

清华大学校长

CONGRATULATION

In 2001, on the occasion of 90th anniversary of Tsinghua University, Mr. Li Zhengdao and Mr. Wu Guanzhong initiated the 1st Art and Science International Exhibition (Symposium). Considerably, this event demonstrated the interaction and integration between the beauty of art and science. The art and science, being a profound and open proposition, has attracted more scholars to plunge themselves deeply into exploration, research and practice.

Both the art and science, though different in perspective and method, aim to explore the truth. We often say that the science seeks for truth, the humanity seeks for goodness and the art seeks for beauty, but in fact, scientificity does exist in art and artistry does exist in science, and both are major force to promote the human development of civilization and social progress.

The mutual complementation and reflection of science with the humanity and art also gives unfailing impetus to the cause of personnel training. Being a comprehensive, research and open type university, Tsinghua University constantly promotes the integration of human spirit and natural science, enhances the interaction of the art subjects with science and engineering subjects, advocates the development of cross and innovative research, and encourages students to strengthen the mastery of other subjects while attending professional education, improve the integrated ability and shape the personal integrity.

At the outset of a new century for Tsinghua University, the 3rd Art and Science International Exhibition (Symposium) opened solemnly in early winter of 2012. It will surely develop a broader space for the future integration of art and science in the field of higher education and the promotion of personnel training, academic research and quality of human life. It is sincerely hoped that the art and science international exhibition will stand long term test, take on a new look and contribute its part to the progress of art and science and the exchange of thoughts and cultures in China and foreign countries.

Chen Jining
Chancellor of Tsinghua University

致辞

在北京一年中最美丽的季节，这本蕴含着海内外艺术与科学无限魅力的作品集与您见面了。"艺术与科学国际作品展暨学术研讨会"已走过了十个年头，成功举办两届，这不啻是对海内外科学与艺术相结合形式的最好见证。

应该感谢李政道和吴冠中二位先生。他们以杰出的业界成就和光辉的人格魅力将人类最伟大的作品——科学与艺术兼容一体，他们站在爱因斯坦、赫胥黎等大师的肩膀上，给我们以启迪和思索。正是有了他们的远见卓识，我们才得以在十年内初沐艺术与科学相结合而带来的光华。

应该感谢文化部、中国文联、清华大学等相关单位。十年来，为促进艺术与科学的和谐发展，他们大力协同、孜孜以求、不懈努力，树立了中华民族在新时代文化艺术新的内涵和发展方向。

应该感谢国内外科学家、艺术家为社会生活、科技前进和艺术繁盛作出的不竭贡献。他们关照心灵、执著创新，发现和创作着跨越时空的文化精品，挖掘着人类对真善美的无限追求。

在党和国家领导人的高度重视下，中国科协作为"第三届艺术与科学国际作品展暨学术研讨会"的支持单位，将与社会各界一道，践行科学精神，推动文化发展，不断培育我国现代科学艺术的宝贵财富，赋予其更新鲜、优秀的内涵。

每一个作品都记录了社会发展的历史脉络，每一个作品都承载着人类文明进步的足迹。能予致辞，荣幸之至。

中国科学技术协会书记处书记
中国科学技术馆馆长

CONGRATULATION

In the most beautiful season of Beijing around the year, this works, which contains infinite charm of domestic and foreign art and science, we be presented in front of you. The Art and Science Works International Exhibition Symposium has gone through ten years and has been successfully held for two terms. It can be said to be the best demonstration of the integration between science and art.

We should thank Mr. Tsung-Dao Lee and Mr. Wu Guanzhong. With their outstanding achievements in their respective field and their brilliant charisma, they advocate the combination between the two greatest works of humanity—science and art. They stand on the shoulders of Einstein, Huxley and other masters, giving us inspiration and enlightenment. It is exactly their vision and farsightedness that enable us to bath ourselves for the first time in the glorious light that is brought by the combination of art and science in the past decade.

We should thank the Ministry of Culture, China Federation of Literary and Art Circles and Tsinghua University among others. For the past decade, they cooperate energetically and work diligently and persistently to promote the harmonious development of art and science, thus have established new connotation and orientation for the development of culture and art of the Chinese nation in this new era.

We should thank scientists and artists both at home and abroad for their inexhaustible contribution to the improvement of social lives, progress of science and technology and prosperity of art. They take care of the soul of mankind, persevere in pursuing innovation, discover and create exquisite cultural works that can transcend the boundary of time and space. Their hard efforts have represented the infinite pursuit of mankind toward truth, kindness and beauty.

As a supporter of the 3rd Art and Science Works International Exhibition (Symposium), we China Association for Science and Technology will work together with all walks of life to practice the spirit of science, promote cultural development, keep cultivating the precious wealth of China's modern science and the art and imbue them with more fresh and more excellent connotations.

Each works have recorded the historical development of human society, and each works demonstrates the footprints of the progress of human civilization. I am very privileged to give these remarks.

Xu Yanhao
Secretary of the Secretariat of China Association for Science and Technology
Curator of China Science and Technology Museum

序一

《第三届艺术与科学国际作品展作品集》顺利出版，我感到由衷欣慰。

吴冠中先生曾说："科学揭示宇宙的奥妙，艺术揭示情感的奥妙。"而李政道先生认为："科学和艺术的共同基础是人类的创造力，它们追求的目标都是真理的普遍性"。在我看来，艺术与科学是一个充满魅力的命题，同时也是一个时代的话题。清华大学举办"艺术与科学国际作品展"，正是顺应历史发展趋势和要求。

清华大学综合性的学科优势、百年历史积淀下的深厚人文底蕴、美术学院雄厚的艺术基础，这些优越条件为科学与艺术的融合互动，构建了独有的创新平台。更重要的是，艺术与科学的互补互动将为清华大学培养拔尖、创新人才提供丰厚的土壤。艺术不仅具有审美功能，还能够帮助学生们超越现实的局限，激发想象力，把不同学科、不同领域结合在一起，这种归纳融合的能力对学生能力的提高是至关重要的。艺术与科学将为学生们提供多种养分，为未来的创新奠基。

2001、2006年，清华大学先后成功地举办了两届"艺术与科学国际作品展暨学术研讨会"，其间，不仅为公众展示了高水平的作品，还通过研讨会的形式汇集了一大批艺术与科学领域的论文，作品与论文交相辉映，成果丰硕，为推动艺术与科学的研究向纵深发展留下了宝贵的财富。本届活动的主题是"信息•生态•智慧"，这是一个既切合时代发展脉搏又充满吸引力的主题，在这样的主题下汇集了国内外艺术家、科学家的众多精彩作品。我们在这里结集出版，既是对本届展览的记录，也希望更多的人能够从这里获得灵感和启迪，我想这才是展览的终极意义。

感谢各支持单位的指导，感谢中国科技馆的鼎力支持、各协办单位的热情扶持，感谢所有为展览、研讨会及系列活动付出智慧和心血的人们。

谢维和

清华大学副校长

FOREWORD

I sincerely rejoice over the successful publishing of *A Works Collection of the 3rd Art and Science Works International Exhibition*.

Mr. Wu Guanzhong said: "The science reveals mysteries of the universe and the art reveals mysteries of the feeling". Mr. Li Zhengdao once puts: "The science and art are commonly based on the human creativity, and seek for the universality of truth". In my mind, the art and science is an attractive proposition, and also a topic of the era. So, the art and science international exhibition hosted by Tsinghua University follows the tendency and demands of historical development.

The advantages, including the comprehensive disciplinary supremacy and profound humanism environment marinating in the history of a hundred years of Tsinghua University and the basis of art of the Academy of Arts & Design, build up a unique innovative platform for the integration and interaction of the science and art. More importantly, the mutual complementation and interaction of the art and science provide fertile soil for Tsinghua University to bring up first ace innovative talents. The art not only functions with the aesthetical features but also helps students to surpass the limits of reality, fire up the imagination and integrate different disciplines and fields. This inductive integration is of great importance for students to promote their capabilities. The art and science will feed students with various nutrients and lay a good foundation for the future innovation.

The Art and Science International Exhibition (Symposium) has been successfully held twice by Tsinghua University, respectively in 2001 and 2006. It not only showed high level works for the public but also collected a great number of dissertations in art and science field through symposiums. These works and dissertations contrasted with each other and contributed abundant achievements. They left valuable wealth for promoting art and science research to develop in depth. The theme of 3rd Art and Science Works International Exhibition (Symposium) is the "Information · Ecology · Wisdom". It is in line with the development pulse of times and full of attraction. With this theme, a great number of masterpieces of foreign and domestic artists and scientists were collected. The collection published is not just a record of this event, but with a hope more people can draw inspiration and capture enlightenment from it. In my mind, this is the ultimate purpose of the exhibition.

We are extremely grateful to all supporters for their guidance, especially for the great support of the China Science and Technology Museum and co-organizers. All the wisdom and efforts paid for exhibition, symposium and series activities will be highly appreciated.

Xie Weihe
Vice Chancellor of Tsinghua University

前言

2012年11月，我们迎来《第三届艺术与科学国际作品展作品集》的出版，这是继2001、2006年《艺术与科学国际作品展作品集》之后艺术与科学展成果的又一次集中展现。

"第三届艺术与科学国际作品展暨学术研讨会"由中华人民共和国文化部、中国科学技术协会、中国文学艺术界联合会支持，清华大学、中国科学技术馆主办，清华大学艺术与科学研究中心、清华大学美术学院、如意科技集团共同承办，2012年11月1日—30日在中国科学技术馆举行。

本届活动主题为"信息·生态·智慧"，旨在汇聚当代国际艺术与科学领域最前沿的研究成果，集合当下艺术与科学领域多元化的认识理念、多样化的表现形式，从信息科学、生命科学和生态科学视角，以艺术美学、生物信息技术和生态智慧为载体，体现人文关怀精神，用新的思路、方法探索未知，创造未来。本届作品展面向全球征集，参展作品来自美国、德国、荷兰、奥地利、法国、西班牙、英国、澳大利亚、意大利、加拿大、厄瓜多尔、丹麦、墨西哥以及中国国内多所艺术院校、研究机构等。参展作品共120余件，包括新媒体艺术、产品设计、建筑与环境设计、视觉传达设计以及艺术创作等体现艺术与科学相结合的科学作品。丰富多样的艺术形式充分体现了科学与艺术相互交融，应和着人类文明的发展进程，共同探寻艺术与科学的理想目标，以视觉的方式展示艺术与科学创造的精神潜质和最新成果。在举办高规格、高水准的作品展览的同时，我们还将举行一系列围绕"信息·生态·智慧"主题的高端研讨会，邀请来自国内外的艺术家、科学家、设计师等就相关研究课题发表演讲、展开交流。

秉承求真、求善、求美的精神，我们致力于将"第三届艺术与科学国际作品展"办成亚洲最具世界水平、具有较高学术价值和前瞻的高端"艺术与科学"展览活动。希望这本作品集能够带给大家更多、更广泛的融合性体验、艺术享受、精神思考和思想启迪。

中国文学艺术界联合会副主席
中央文史研究馆副馆长
清华大学美术学院名誉院长

PREFACE

In November, 2012, *A Works Collection of the 3rd Art and Science Works International Exhibition* will be published. Successive to collections in 2001 and 2006, it is another central achievements display of the art and science exhibition.

Supported by the Ministry of Culture of the People's Republic of China, the China Association for Science and Technology, and the China Federation of Literary and Art Circles, the 3rd Art and Science Works International Exhibition (Symposium) is hosted by Tsinghua University and the China Science and Technology Museum and co-organized by the Art & Science Research Center of Tsinghua University, the Academy of Arts & Design of Tsinghua University, and Ruyi Science & Technology Group. It will be held in the China Science and Technology Museum from November 1st, 2012 to November 30th, 2012.

With the theme of "Information · Ecology · Wisdom", this event assembles the latest cutting-edge research results in the field of art and science, integrates the current diversified ideas and diversified art forms, embodies the spirit of humane care, and attempts to explore and create the future through new thoughts and methods by adopting the perspectives of information science, life science and ecological science and making art aesthetics, biological information technology and ecological wisdom as the carrier. The Exhibition solicits works from all over the world. The works on display come from the art colleges and research institutes in USA, Germany, Netherlands, Austria, France, Spain, UK, Australia, Italy, Canada, Ecuador, Denmark, Mexico and China. Totally, this event assembles over 120 pieces of works in a variety of forms, including new media art, product design, architectural and environmental design, visual communication design, and artistic creations featuring the combination of art and science. These works in a variety of art forms fully reflect the integration of science and art, jointly explore the ideal goals of art and science in combination with the development history of human civilization, and display the spiritual potentials and latest achievements created by art and science in the visual forms. Besides the high-class and high-level exhibition, a series of high-end symposiums focusing on the theme of "Information · Ecology · Wisdom" will also be held, in which the domestic and overseas artists, scientists, designers and other experts will deliver speeches and develop exchanges on relevant research subjects.

In compliance with the spirit of seeking truth, goodness and beauty, we are dedicated to build the 3rd Art and Science Works International Exhibition into a high-end art and science exhibition featuring world level and superior academic interest and prospects in Asia. It is hoped that the collection may come up with more and broader integrated experience, enjoyment of art, spiritual thinking and enlightenment of thoughts.

Feng Yuan
China Federation of Literary and Art Circles Vice Chairman
Deputy Director of Central Research Institute of Culture and History and Honorary
Honorary Dean of the Academy of Arts and Design of Tsinghua University

目录
CONTENTS

组织机构	004	ORGANIZATION
致辞	008	CONGRATULATION
序	012	FOREWORD
前言	014	PREFACE
主题阐释	018	EXHIBITION THEME
作品	020	WORKS
后记	192	POSTSCRIPT

主题阐释

信息 · 生态 · 智慧

信息时代日新月异的科学、技术推动人类社会自身的进步;生态文明昭示着人类生存方式的可持续发展;人类智慧以创新的价值实现崇高的人文理想。

展览集合当下艺术与科学领域多元化的认识理念、多样化的表现形式,从信息科学、生命科学和生态科学视角,以艺术美学、生物信息技术和生态智慧为载体,关注人类终极理想和精神,用新的思路、方法探索未知、创造未来。

EXHIBITION THEME

Information · Ecology · Wisdom

Fast developing science and technology in this era of information promotes development of human society; ecological civilization announces sustainable evolution of human lifestyle; human wisdom realizes sublime humanistic ideal with the value of innovation.

The exhibition integrates the current diversified ideas and diversified art forms, adopts perspectives of information science, life science and ecological science, makes art aesthetics, biological information technology and ecological wisdom the carrier, pays attention to ultimate ideal and spirit of human being, and attempts to explore the unknown and create the future through new thoughts and methods.

作品

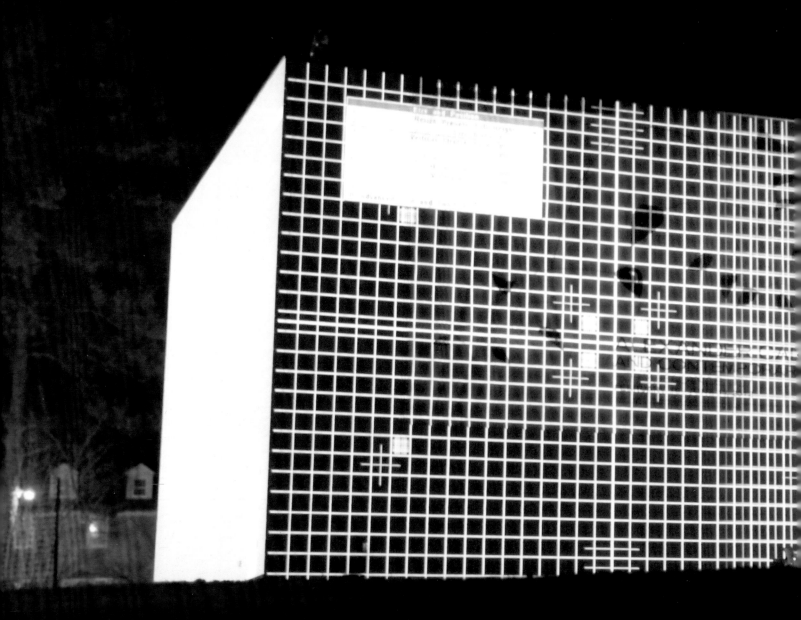

WORKS

多感的建筑（虚拟作品），2012
比尔·希曼、托德·贝雷思（美国）
计算机、投影机、音频系统等
© 比尔希曼，托德贝雷思（杜克大学）

A China of Many Senses (VR), 2012
Bill Seaman & Todd Berreth (USA)
Computer, Projectors, Audio System
© Bill Seaman & Todd Berreth (Duke University)

《多感的建筑》用一套生成式计算系统来探究编程创造的艺术魅力。作品运用大量的模型数据库、视频资料、数字静态影像、声频资料以及文本资料，并用作者专有的代码（以C++/OpenGL/OpenFrameworks写成），以愈益智能化的方式来加以拼贴和整合。

"A China of Many Senses" explores aspects of programmed machnic creativity via a generative emergent computational system. The work draws on an extensive database of specific models, video materials, digital stills, audio materials, and textual materials that are combined and recombined in an ongoing intelligent manner, via authored proprietary code (written in C++/OpenGL/OpenFrameworks).

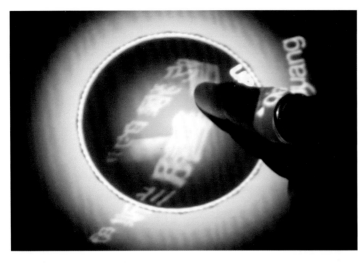
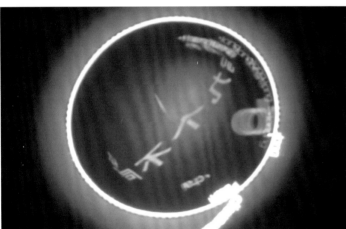

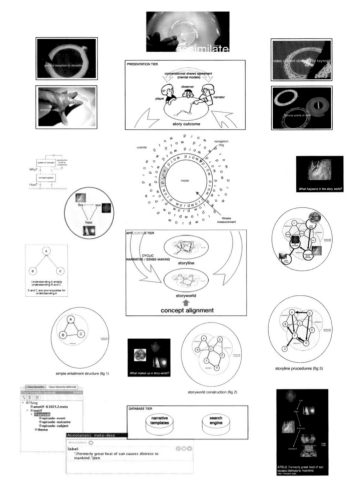

吸收——协作叙事构建，2012
丹明·荷尔斯（澳大利亚）
电脑、投影机、微软 Kinect 或 Touch Table 界面（例如 Surface 界面）
© 丹明·荷尔斯

Assimilate —Collaborative Narrative Construction, 2012
Damian Hills (Australia)
PC, Projector, Microsoft Kinect or Touch Table Interface (eg Surface)
© Damian Hills

《吸收——协作叙事构建》是一个让参与者可在三维视觉空间中构建视觉叙事的、在线的协作环境。运用手势界面，2—4名参与者可输入关键词，通过互联网搜索获得在线媒体协作叙述一个进行中的故事。该系统是一个让参与者的搜索标准与取自神话和民俗数据库的模板故事相匹配的可视化搜索引擎。参与者对叙事结果及其相关行为进行选择，同时搜索结果将自动生成为可视化风格。娱乐化的界面操作促进了角色扮演和相互对话，其意义和内涵是通过一个持续不断得到搜索反馈和叙事模板可选性而循环的过程。

此项目是作为创造力和认知工作室（CCS）实践导向博士研究项目的一部分而开发的。CCS是一个为推进数字媒体和艺术实践的发展和研究而成立的国际认可多学科研究中心，为得到本国和国际承认的艺术家和科研人员提供一个空间，这个空间作为平等的合作伙伴在实践导向研究中相互协作、实验和创造。

CCS致力于通过出版物、展览、举办国际会议，以及提供高质量的研究生教育课程等方式，在国际间广泛传播其研究成果。

The "Assimilate" project is an online collaborative environment that allows participants to visually construct narratives in a 3D virtual space. Using a gestural interface, two to four participants can collaboratively narrate an ongoing story using online media obtained through an internet keyword search. The system is a visual search engine that aligns the participant's search criteria with template stories drawn from a database of mythology and folklore. The search results are styled into generative behaviors that visually self-organise while participants make choices about the narrative outcomes and their associated behaviors. The playful interface promotes conversation and role-playing as meaning and connotation are cycled through a continuous process of search result feedback and narrative template selection.

This project was developed as part of a practice based research PhD with the Creativity and Cognition Studios (CCS), an internationally recognized multi-disciplinary environment for the advancement and understanding of practice in digital media and the arts. It provides nationally and internationally recognized artists and technologists with a space in which to collaborate, experiment and create, as equal partners, in practice-based research.

CCS is committed to disseminating its results internationally through research publications, exhibitions, the co-ordination of an international conference series and through the provision of high quality postgraduate education.

自古至今，人类在"以人为本"的思维下，不断地对外界进行改造，人类通常利用自然界的规律，来满足自身的需求，而这一切，往往不是源自"物体"内在的本意。

在信息时代下，日新月异的数字技术，使人类可以模拟自然现象，使虚拟物体在数字环境里具有类似自然的"自我演化"能力。

这个项目的目的在于探讨这种可能性，OBJECT#SQN1-F2A 是第一件从概念到现实的作品，由 146 个面组成，三维数据来自设计师自己，完全按照个人形态生成制作，为了尊重物体生成的结果．制作过程中尽可能减少人对形态的干预，以突出其最原始的属性，从而体现出一种真诚、自然之感。

Man is constantly transforming the outside world under the thinking of "people paramount" since the ancient times to date, and usually man meets their own needs by making use of the natural laws, which is not the intrinsic own intention of the "substances".

The constantly changing digital technology in the information times can enable man to imitate natural phenomena, and bring about capacity of "self evolution" similar to natural capacity to virtual objects in digital environment. The project is intended to explore such possibility. OBJECT#SQN1-F2A is the first works that comes from concept to reality, and is composed of 146 surfaces. The 3-D data come from the designer, and is made in full accordance with the individual forms. Human intervention is reduced as much as possible during production to respect the results generated, and highlight the most primitive properties so as to embody a sincere and natural sense of beauty.

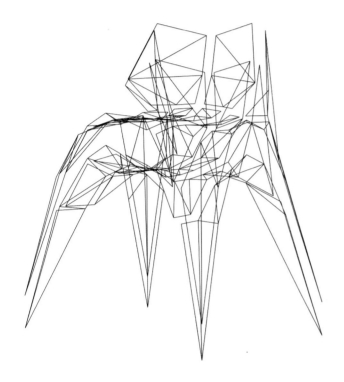

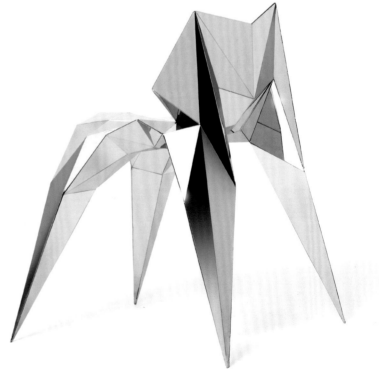

数字智造 OBJECT#SQN1-F2A, 2010
张周捷（中国）
不锈钢
© 张周捷数字实验室

Digital Intelligent Creation OBJECT#SQN1-F2A, 2010
Zhang Zhoujie (China)
Stainless Steel
© Zhang Zhoujie Digital Lab

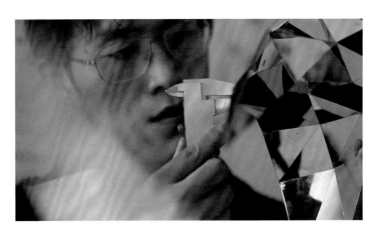

《迪斯科 Pada》是一款互动音频游戏，在数字游戏《乓》的基础上开发而成。这款游戏仅使用音频输出，作为对我们当今环境中大量的可视信息的反应。

它用音乐取代了《乓》中的乒乓球，引导玩家进入房间中的正确位置来拦截"移动的音乐"。佩戴上无线耳机，玩家根据"乒乓球"的位置以及自身的移动而听到音乐在环境中移动。《迪斯科 Pada》测量玩家所处的空间位置。在较大的地方需要玩家奔跑；而在狭小的空间中则需要微小移动。

一旦玩家进入正确的位置，音乐声就靠近玩家面前随后被反弹到对手那里，这时候音乐中就会添加一种器乐声，作为对玩家的奖励。如果玩家位置不正确，则音乐会从玩家身旁飘过，而且会减去一种器乐。这套游戏的目标就是要在歌声结束之前获取并保住全部的音乐层。

"Disco Pada" is a cooperative audio game based on the arcade game "Pong", wherein digital Ping-Pong is played. As reaction on the amount of visual information in our present environment, this game only uses audio as output. With the focus on generation, the game is flexible and expressive. The ball is replaced by music that will guide the players to the right location in the room in order to block "the moving music". This is done by giving feedback through wireless headphones and using spatialized sound. This way the players hear the music moving through the environment according to the location of "the ball" and their own movement. "Disco Pada" measures the player's position relative to the size of the space he/she is in. This creates a different game experience in different environments. Where a large area requires running; a confined space require subtle movements in order to play the game.

Once a player is on the right location, the music sounds as if approaching directly from the front and instead of passing the player, the music will bounce back to the other player. This success is rewarded by an extra layer of instruments in the music. If one of the players fails, a layer will be removed. The goal of the game is to obtain and preserve all the music layers before the song has ended.

迪斯科 Pada,2012
刘伟(中国)、
Floor-Jan van Schaik(荷兰)、
Maartje van den Hurk(荷兰)、
Pascale de Rond(荷兰)、
Hu Mengfei(中国)、
Tommie Varekamp(荷兰)
混合材料
© 刘伟

Disco Pada, 2012
Liu Wei (China)
Floor-Jan van Schaik (Netherlands),
Maartje van den Hurk (Netherlands),
Pascale de Rond (Netherlands),
Hu Mengfei (China),
Tommie Varekamp (Netherlands)
Mixed Material
© Liu Wei

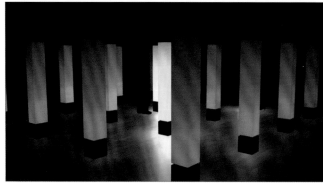

置换共鸣，2012
约翰·费尔沃克（美国）
混合材料
© 约翰·费尔沃克

Displaced Resonance, 2012
John Fillwalk (USA)
Mixed Material
© John Fillwalk

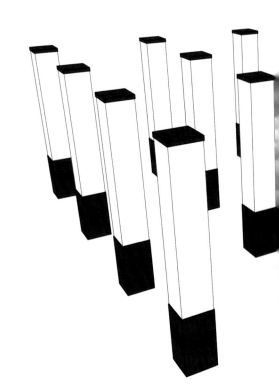

这件作品包括十六个雕塑单体,它们依据参观者距离的远近控制照明和声音的响应。内置全频扬声器让声音环绕这些雕塑。配有红外摄像头的电脑系统跟踪参观者的移动,从而控制声音的分布和相关的 LED 照明。这件作品的名称指的是把声音从它们原来的物理环境和文化语境中剥离开来,然后置换新的物理共鸣和文化共鸣。

这件作品创作过程中曾使用实物模型和虚拟模型来设计并制作雕塑形状、产生互动功能。实物原型在虚拟模型中被解读,考察雕塑与电脑化身之间的空间交互——雕塑会与虚拟化身的出现和靠近进行互动。这种互动功能被编写在虚拟世界里,然后再次以可感知的方式在现实中以实物来加以模仿。经过几次虚拟迭代后,雕塑形状新近被重新解读,然后被制作成实物形状,在公共场合中进行互动。在将来的迭代中,《置换共鸣》再一次的实物表达中将具备另一项能力,能够受到远程虚拟互动的影响。

This installation consists of sixteen sculpture forms, networked in a responsive grid controlling light and sound, based on visitor proximities. Full-range internal loudspeakers drive sound through the sculptures. A computer system using an infrared camera, tracks the movement of visitors and responds by controlling the distribution of sound to the tubes, while also controlling LED lighting associated with the sculptural forms. The title of the installation refers to removal of the sounds from their original physical and cultural contexts, and the imposition of new physical and cultural resonance. The protyping process included both physical and virtual models to design and build both the form and the interaction. The physical prototypes were interpreted in a virtual model, investigating the spatial interaction of the structure with an avatar—where the sculpture interacted with avatar proximity and presence. The interactive functionality was scripted in the virtual world and again modeled in the sensored version in physical reality. After several virtual iterations, the form was recently re-interpreted and fabricated to have a physical form and be interacted with in a public context. In future iterations, "Displaced Resonance" in its once-again physical manifestation has the additional capability now to be influenced by remote and virtual gestures and interactions.

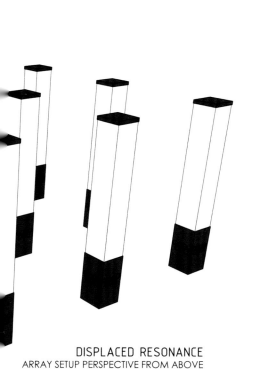

DISPLACED RESONANCE
ARRAY SETUP PERSPECTIVE FROM ABOVE

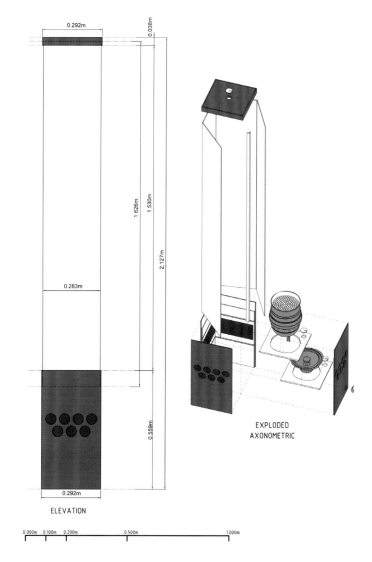

ELEVATION

EXPLODED AXONOMETRIC

互动作品《能流》的参与者可以根据他们的社会关系和空间关系来创建路径，并绘制于虚拟三维环境之中，参观者相互靠近便创造出节点。

Participants in "CONIFLUENCE" create pathways based on reactions to both social and spatial relationships. There is a virtual response and persistence to interactions, as the routes are drawn and sculpted in the three-dimensional environment—forming nodes that are created via visitors' proximity.

能流，2012
John Fillwalk、Michael Olson、艺术家与 IDIA 实验室（美国）
投影仪、幕布、电脑、扬声器
© 约翰·费尔沃克

ConIFLUENCE, 2012
John Fillwalk with Michael Olson, Composer and IDIA Lab.(USA)
Projector, Screen, Computer and Speakers
© John Fillwalk

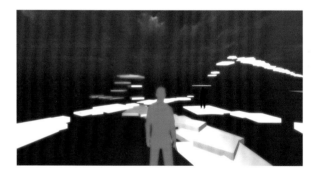

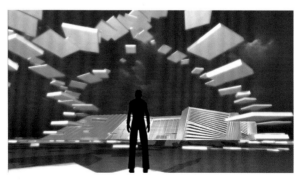

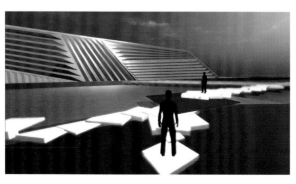

作品《FLICKR ™ GETTR v6》将照片分享网站 Flickr ™ 与虚拟博物馆环境连接起来，让参观者可通过输入搜索关键词，将 Flickr ™ 网站上的相关照片引入虚拟环境之中，从而创建出一个动态的虚拟空间。

"FLICKR™ GETTR v6" connects the social image web service of Flickr™ to the virtual museum environment, allowing visitors to create a dynamic cloud of spatial imagery by entering a search term of their choice, that pulls in related images from Flickr™ into the virtual environment.

FLICKR™ GETTR v6，2012
John Fillwalk、Jesse Allison、艺术家与 IDIA 实验室（美国）
投影仪、幕布、电脑、扬声器
© 约翰·费尔沃克

FLICKR™ GETTR v6, 2012
John Fillwalk with Jesse Allison, Composer and IDIA Lab, (USA)
Projector, Screen, Computer and Speakers
© John Fillwalk

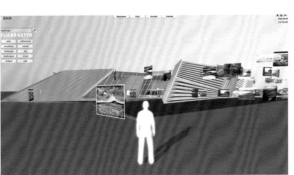

虚拟装置艺术作品《仿真》，让参观者体验虚拟博物馆的形体和音效。该作品在虚拟空间中逐步成型，最终形成虚拟建筑的结构。如果虚拟博物馆环境中有多名用户，他们的虚拟化身会互相作用，合作创造出绘画和雕塑。

In the virtual installation, "Proxy", visitors shape the construction of a sculptural and sonic response to the virtual museum. The work progresses to form, eventually transforming to become structural support for the building. When multiple users are in the environment, their avatars interact with one another to create collaborative painting and sculpture.

仿真，2012
John Fillwalk、Michael Olson、艺术家与 IDIA 实验室（美国）
投影仪、幕布、电脑、扬声器
© 约翰·费尔沃克

Proxy, 2012
John Fillwalk with Michael Olson, Composer and IDIA Lab. (USA)
Projector, Screen, Computer and Speakers
© John Fillwalk

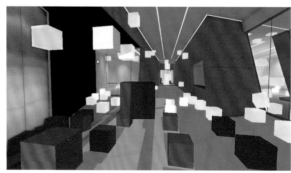

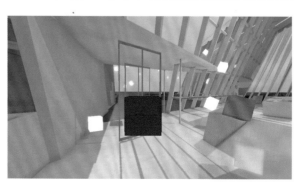

《北京勘测》使用中国北京市实际地点的实时气象数据,构建出一个引人入胜的仿真景观。诸如旗帜、光线、时间、风和云的仿真图像根据真实的气象数据反映到虚拟景观之上。

"Survey for Beijing" is an immersive landscape simulation using real time weather data from the physical location in Beijing, China. Representations of surveyor's flags, light, time of day, wind and clouds are superimposed onto the virtual landscape in accordance with real-life weather data.

北京勘测,2012
John Fillwalk、Keith Kothman、艺术家与 IDIA 实验室(美国)
投影仪、幕布、电脑、扬声器
© 约翰·费尔沃克

Survey for Beijing, 2012
John Fillwalk with Keith Kothman, Composer and IDIA Lab. (USA)
Projector, Screen, Computer and Speakers
© John Fillwalk

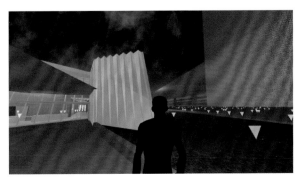

头脑风暴，2012
李维、朱鼎亮、胡俊铭（中国）
水晶、EEG脑电波感应器、投影、电机
© 李维、朱鼎亮、胡俊铭

Brainstorm, 2012
Eric Lee, Martin Zhu, Jimmy Hu (China)
Crystal, EEG Brainwaves Inductor, Projector, Electric Motor
© Eric Lee, Martin Zhu, Jimmy Hu

单凭人脑的念力去控制实物，是全人类共同的梦想。脑电波控制技术将这个梦想变成了现实。《头脑风暴》就是通过脑电波控制技术与物联网技术，使悬挂在天顶电机阵列下的各种实物，在参与者的脑电波操控下在空间中升降聚散。同时在光的投影下，其重新组合的形体，形成了一个新的符号形象，使观众与参与者获得了一种仿如自己的思维意识从模糊散乱到逻辑清晰的过程。

It is the common dream of man to control real objects with merely the mental power of brains. The controlling technology of brainwaves has turned the dream into reality. "Brainstorm" causes the various real objects suspended under the array of electric motor on the ceiling to rise, fall, gather and scatter in the space under the control by the brainwaves of the participants. Meanwhile, the newly-combined body forms a new symbolic image under the projection of light, bringing about the feeling among the audience and the participants that their own thinking and awareness have transited from dim and disorderly to logic and clear.

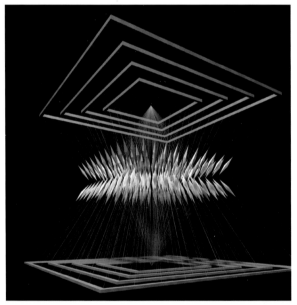
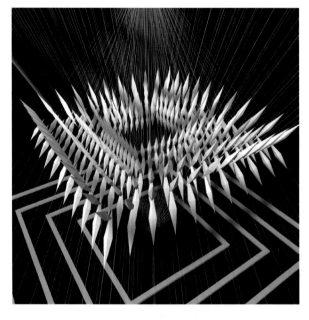

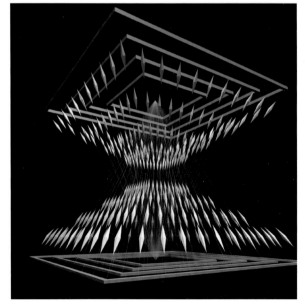

《针鼹》是一件互动式有声雕塑，它就像一只大惊小怪、连滚带爬的生物，有自己的声音，会"吱吱"地尖叫，还会对人们的出现作出反应。它是用上色的铁丝做成的，坐在内置了电子元件的底座上，如同一幅乱糟糟的线图被注入了生命，也被赋予了外形，并且创造出一片磁场。它可以测量到环境中的静电变化，通过与一套专门设计的相锁回路系统驱动扬声器。当人们触碰并干扰了它周围的电磁场时，这件雕塑就会发出声音。

Interactive Sound Sculpture, "Echidna" is like a fussy tumbled creature that has its own (electronic) voice—it will squeak and react to human presence. It is made of coloured wire, like a messy line drawing infused with life and shaped to create a magnetic field. It sits on a plinth with electronics inside. The works combines a circuit, which directly measures electrostatic changes in the environment and this together with a custom designed phase locked loop system is used to drive an audio speaker. When you touch and disturb the electromagnetic field around it, a sound emerges from the sculpture.

针鼹，2010
蒂娜·贝赫（丹麦）
混合材料
© 蒂娜·贝赫

Echidna, 2010
Tine Bech (Danmark)
Mixed Material
© Tine Bech

该作品将以动态图形和交互方式呈现城市信息生态系统的变化。作品构建了一组传达城市信息的空间交互装置，结合多屏的、叙事的、反射性的界面形态，呈现出网络时代城市的生态特征。作品通过数据挖掘和信息可视化的手段，以讲故事的方式来反思城市化面临的问题。其中的动态信息内容将以城市感知、生活归属、生态持续三个相互连接的主题进行呈现。城市感知是通过网络结构展现城市信息的动态变化；生活归属是市民与城市之间的互动与包容；生态持续则是环境与人的活动之间的平衡关系。该作品既以动态信息流呈现信息时代的城市变化，亦以多视角的活动映像展现了对城市生态演化的反思。

The works demonstrates the changes of urban information and ecological system through dynamic infographics and interactive modes. The interactive installation displays the images of cities in the network age by a group of modules with multi-screen, reflective and narrative interfaces. It reflects on the problems in the process of urbanization by means of data mining, information visualisation and interactive storytelling. The contents of the dynamic information therein will be displayed in three interlinked themes such as urban sensing, living belonging and ecological sustainability. Urban sensing is to visualize the dynamic changes of urban information through a network structure, while living belonging is the interaction and inclusive pattern between urban residents and the city; ecological sustainability is the balanced relationship model between environment and human activities. The works has not only represents the changes of smart city with dynamic information flow, but also the reflection toward urban ecological evolution through images of activities from multiple perspectives.

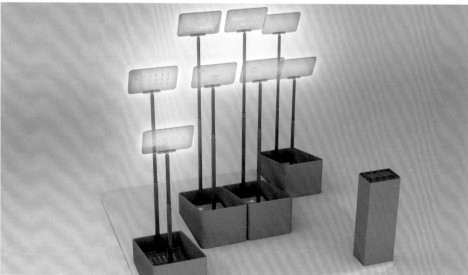

城市信息流，2012
付志勇（中国）
动态数字媒体、交互装置
© 付志勇

Eco-City Flow, 2012
Fu Zhiyong (China)
Dynamic Digital Media, Interactive Device
© Fu Zhiyong

这是一个互动艺术装置，通过分析观众大脑的α波和β波实现互动：当观众戴上头带后，一只蝴蝶将投影在面前的窗棂中，并按照观众的脑电信号，在竹影中上下飞舞。观众可以通过放松精神来控制蝴蝶的飞行轨迹，让蝴蝶穿过层层密林。观众的注意力集中程度、眨眼频率也会对画面产生影响。一旦蝴蝶到达终点，一切又将散为虚无。庄周梦蝶，焉知是梦？

This is an interactive installation driven by mind waves. Putting on the headset, the audience will see a shadow of a butterfly projected on the front window. A player could control the movement of the insect by his/her meditation. Once the butterfly reaches the end of the bamboo woods, all illusions in the window will disappear. As Zhuangzi, an influential Chinese philosopher dreamt and described, how could you know it was a dream?

空窗子, 2012
黄石、李敬峰（中国）
背投投影装置、脑电波感应器
© 黄石、李敬峰

Empty Window, 2012
Huang Shi, Li Jingfeng (China)
Projector, Brain Wave Sensors
© Huang Shi, Li Jingfeng

150°的旋转台上安装摄像机，其拍摄的映像用全景和鱼眼的两种方式变换，通过"鱼眼用向月乐器台"和"全景乐器台"两个台的配合，使其映像通过互动展示风景乐谱二重奏，现场可让观众参与作品。该作品试图展现池塘中的金鱼眼中世界的样子。

The 150 degrees of rotation installation on the camera, the film image with panoramic and fish eye two ways to transform, through the "fish eye to use on instruments table" and "panoramic instruments table "two tables to cooperate, make its image through the interactive display scenery music duet. Field can let the audience to participate in work.

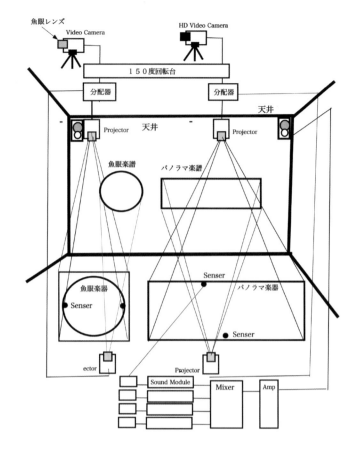

鱼眼舞，2012
山本圭吾（日本）
混合材料
© 山本圭吾

Fish Eye Dance, 2012
Keigo YAMAMOTO (Japan)
Mixed Material
©Keigo YAMAMOTO

将在场的距离感打碎，就能用本来毫无干连的图片构成一幅具有新的统一视点的拼贴画。要制作一幅拼贴画，首先要打好透视网格，然后再衬上匹配的图像片段。首先要从可能匹配的海量图片中挑选合适的素材，之后摒弃边框并加以修改。通过定义描述照片原件内容的额外特质，就能影响图像片段的选择结果，而最终的拼贴画正是由这些片段构成的。如此一来，与原始素材之间形成一种语义上的关联，使拼贴平添了一种语境。

Using the extracted image segments, it is now possible to form collages of originally different pictures with a new common perspective. In order to compose a collage, a perspective-grid is defined and a lining of matching image segments is being applied. The segments are not altered to match the frame but fitting ones are chosen from the sheer mass of possible pieces. By defining additional keywords which describe the content of the original photographs, the selection of segments used for the final composition can be influenced. Thus a contextual layer is added through the semantic linking with the source material.

在场的凝聚，2012
托尔斯滕·博塞尔特、本雅明·茅斯（德国）
数码设计
© 托尔斯滕·博塞尔特、本雅明·茅斯

Extracts of Local Distance, 2012
Torsten Posselt, Benjamin Maus (Germany)
Digital Design
© Torsten Posselt, Benjamin Maus

《点燃我的激情》是为博物馆或临时展览而构思的一套交互式装置，它的交互原则非常简单、直观。只要使用一根简单的火柴，参观者就能"点燃"投射的画面。画面在观者头顶生成，呈现给在场的所有参观者。火柴的火焰是动画的起始点，每一幅画面都配有不同的声音，这使整个场面更加盛大。这套装置不仅具有很强的娱乐性，也是对互动与可视方式的研究。

"Light My Fire" is an interactive installation thought for the museum context or temporary exhibitions. The interactive principle is very simple and intuitive.
With the help of a simple match, the visitor is able to "Fire" projected animations that appear over his head and are visible for all other visitors. The flame of the match is used as the starting point for the animations. Every animation is supported by a sound, giving a more spectacular sensation. This installation is playful but at the same time a research in the interactive and visual approach.

点燃我的激情，2008 年完成 1.0 版
弗洛里安·皮泰、埃里克·莫齐耶（瑞士）
电脑、外接绘图板、视频放映机、视频电缆、红外线摄像、自制火柴棍、扩声器、放大器、低音炮、音频电缆
© 洛桑艺术设计大学西格马希克斯实验室

Light My Fire, 2008 for version 1.0
Florian Pittet, Eric Morzier (Switzerland)
Computer, External Graphic Card, Video Projectors, Video Cable, IR Camera + Support Matches, Home Made Strike Bar, Loud Speakers, Amplifier, Sub Woofer, Audio Cable.
© SIGMASIX | ECAL

中间黑盒子里放一个蜡烛，周围的传感器可以感知烛光随风的偏向角度，带动四周的 LED 灯的明暗互动。

A candle is placed in a black box in the middle, and the sensors around can sense the obliquity of the candlelight with the wind, and leads to the interactive lighting and shading of the LED lights around.

光 & 烛，2012
牛淼（中国）
综合材料交互装置
© 牛淼

Light & Candle，2012
Niu Miao (China)
Comprehensive Materials, Interactive Device
© Niu Miao

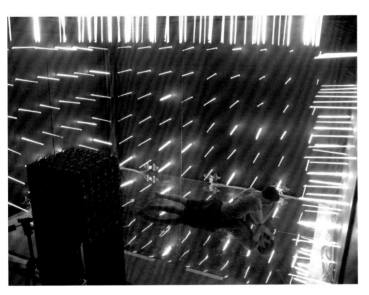

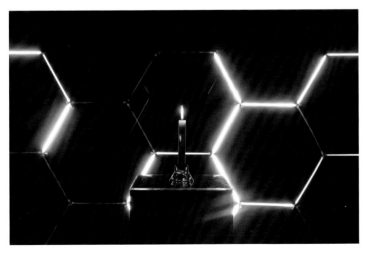

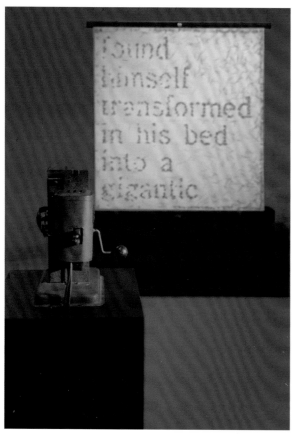

逃离, 2012
洛朗·米格侬努、克里斯塔·索默雷（法国/奥地利）
交互式装置
© 为瑞士"视图——当代艺术空间"创作

Escape, 2012
Laurent Mignonneau, Christa Sommerer (Fance/Austria)
Interactive Installation
© Developed for The View Contemporary Art Space, Switzerland

《逃离》这件作品探讨尝试飞行的问题，最早是作为瑞士萨伦斯坦市民防空洞专用的装置而研制的，于索默雷和米格侬努的"视图——当代艺术空间"平台上进行专场展出。

这套作品包括20世纪40年代一台过时的胶片投影机和一张过时的投影屏幕。投影机经过了改装，使之能够容纳一台现代的小型数字投影仪，并支持一些传感技术。当参观者进入房间后，会看到一只苍蝇停在投影屏幕表面上。当大家开始转动老投影机的手柄时，苍蝇便开始发狂地乱转，就好像即将落入陷阱、正在试图逃离一样。如果继续转动手柄，就会有更多的苍蝇飞来挤成一团，如同它们在屏幕表面上发现了什么养料似的。

某一时刻文本开始从苍蝇群中显现，这时如果继续转动手柄，屏幕上就会出现可以辨识的一段文字——弗朗茨·卡夫卡《变形记》中的一段，讲述主角格里高尔·萨姆沙意识到自己一夜变形，成了一只巨大的昆虫。

The installation "Escape" deals with the issue of flight attempt. It was originally developed as site specific installation in the civilian air raid shelter in Salenstein Switzerland as part of Sommerer and Mignonneau's solo show at the The View Contemporary Art Space.

The installation consists of an antique film projector and an antique projection screen from the 1940s. The projector was modified to hold a small video projector and some sensor technology. When visitors enter the room, they see a fat fly sitting on the projection surface of the screen. Once they start turning the handle of the old projector, the fly starts to frenetically move around, like it would be trapped, trying to escape. When continuing to turn the handle, more and more flies would come, packing themselves together as they would have discovered some nutrition onto the screen surface. At one point text starts to form out of the fly pack and when one continues to turn the handle, a text becomes legible. It is a chapter of "The Metamorphosis" by Franz Kafka where the protagonist Gregor Samsa realizes that he has transformed into a gigantic insect overnight.

理性与情感，逻辑与形象，是构成人类不可分离的思维整体。作品以东西方的两个典型雕塑人物形象为摄影对象，突出体现了两种不同文化认知和艺术表现特征：客观写实与夸张写意——逻辑的与意象的，它们都是人类艺术精髓。

创作者将现代科技与摄影艺术相结合，将相同的摄影对象、不同的聚焦点的两次影像叠加合成为一张。通过编程控制正常光照和特殊光照的轮换，在观者面前，微妙地交替呈现两个不同聚焦效果的影像。观者思维游离在"是"与"似"间。请：静、观、悟。

Sense and sensibility, logic and image, are all the integral parts of the human being's whole thinking. The works selects two typical Oriental and Western sculpture figures as the photographic object, and highlights two different kinds of cultural cognition and characteristics of artistic expression: objective realism and freehand brushwork, namely, logic vs. image, both are essence of art.

Combining the photographic art and the modern technology, the artist overlays two images that reflects the same photographic object but in different focusing points into one. The rotational presentation of the normal lighting and special lighting is controlled via programming, and two images with different focusing effects are delicately represented in front of the viewers on a rotating basis, drifting between "being" and "resemblance". Please feel free to enjoy and meditate.

"是"与"似"，2012
鲁晓波（中国）
照片、艺术微喷、紫外显影
© 鲁晓波

"Being" and "Resemblance", 2012
Lu Xiaobo (China)
Photo, Giclee, Ultraviolet Development
© Lu Xiaobo

作品试图通过黑白的佛教观想壁画和象征心灯的烛光图，使用特殊的两个图层交替在同一个画面中显现的崭新技术手段，生成一种直观的意象效果，尝试让抽象的思想直接诉说于一种艺术形象。《观音曼荼罗》拍摄于西藏阿里的扎达县托林寺。该寺始建于北宋时期（公元996年）是古格王国（公元10—17世纪）在阿里地区建造的第一座佛寺，其建筑风格和壁画明显受尼泊尔和印度的影响，是研究当地古代建筑、雕塑、绘画的珍贵实物。新的信息科技手段为艺术展现宗教智慧与审美魅力提供了新的可能性。

The works attempts to generate a visible image effects and directly express the abstract conceptions in an artistic image by making use of the black-white Buddhist conceptive murals and candlelight picture that symbolizes mental light, and brand-new technical methods by means of two special layers alternating in the same picture. "Avalokitesvara Datura" was taken in Tuolin Temple, Zhada County, Tibet. The temple was first built in Northern Song Dynasty(996 AD), and was the first Buddhist temple built in all region in the Guge Kingdom. Its architectural styles and murals have been obviously influenced by Nepal and India, and are rare real objects to study local ancient buildings, sculptures and paintings. The new information technological methods have provided new possibility to demonstrate religious wisdom and aesthetic charms in art.

 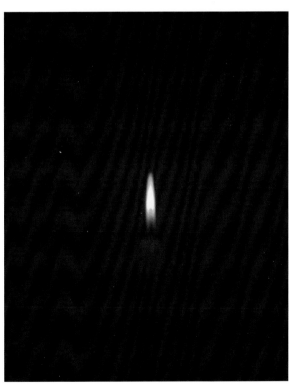

观音曼荼罗，拍摄于2004年，制作于2012年
冯建国（中国）
照片、艺术微喷、紫外显影
© 冯建国

Avalokitesvara Datura, Picture 2004, Made 2012
Feng Jianguo (China)
Photo, Giclee, Ultraviolet Development
© Feng Jianguo

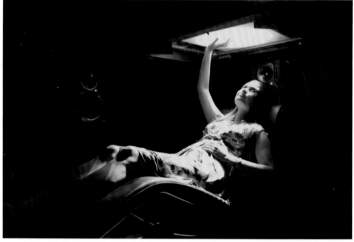
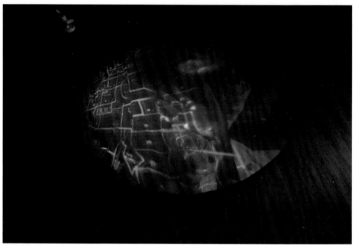

有限,2011—2012
基思·M·阿姆斯特朗(偕同罗杰·迪恩、斯图亚特·劳森及达伦·派克)(澳大利亚)
灵敏透明触摸屏、躺椅、投影机、计算机、电子元件、4声道音响、移动装置、电线、组装件
© 基思·M·阿姆斯特朗

Finitude, 2011-2012
Keith M. Armstrong(With Roger Dean, Stuart Lawson & Darren Pack)(Austrslia)
Touch sensitive transparent screen, Couch, Projector, Computer, Electronics,
4 Channel sound, Mobiles, Electronics, Wiring, Assemblies
© Keith M. Armstrong

《有限》（第三版）是一件新颖的大型媒体／雕塑艺术科学作品。每一位参与者或参观者可以舒适地仰卧在躺椅上，面对着一块半透明的塑胶玻璃屏幕，它可以放映投射的三维图像，并且十分灵敏，能够感应到最轻微的手指触摸。在LED聚光灯的照射下，参观者可以透过这块玻璃屏幕看到悬挂在其上方的装置实物不断移动所产生的动画影像。这些悬挂在上方的装置模型实物在缓慢的旋转中，能够实时构成全息式的"景观"，从视觉上将"真实"与"虚拟"的媒介形式结合起来。

参观者通过轻微触屏互动，在三维图形、投影画面的实物移动以及有如身临其境的四声道音频的共同作用下，实现实物与虚拟媒介之间浑成与融合。同时，房间四周安放了四只扩声器，它们能够传递出丰富的交互式音频，从而在声音和物理两方面影响的作用下，促进参观者与作品之间的互动。

《有限》是作者在澳大利亚野生动物管理局实习期间与澳大利亚内陆桉树丛林地区的生态科学家们共同创作的，该局是在科学基础上从事稀有哺乳动物保护工作的组织。这件作品希望通过表现人类以及许多其他物种的"剩下的时间"，探讨了如今人类是应当把时间赠予未来，还是要从未来取走时间（从生态的角度）的问题。

Finitude (V3) is a major new media/sculptural art-science work. Each participant/viewer lies comfortably on their back. Directly above them is a semi-transparent Perspex screen that displays projected 3D imagery and is also sensitive to the lightest of finger touches. The participant can see through its animated surface to a series of physical mobiles suspended above, lit by LED spotlighting. These mobiles consist of a slowly rotating series of physical models allowing the real time composition of holographic-style "landscapes" that visually combine both these "real" and "virtual" media forms.

Through subtle touch-sensitive interactivity the participant then has influence over both the 3D graphic imagery, the physical movements of the diorama and the 4 channel immersive soundscape, creating a seamless blend of physical and virtual media. Four speakers positioned around the room deliver a rich interactive soundscape that responds both audibly and physically to interactions.

Finitude was created in collaboration with ecological scientists from Australia's Mallee outback country during residences with the Australian Wildlife Conservancy—a rare mammal conservation/science based organisation. The works speaks about "time left" for both ours and many other species and considers how we might now choose to give to, or take time away from, the future (from an ecological perspective).

《闪耀的蜂巢》使用模拟群居昆虫群集模式的计算机算法，将关于动物意识的实时网上对话映射到一件交互式音频—视频装置中，它生成动画效果的文本和声音，描摹 Twitter 或 Sina Weibo 网站上关于动物的谈话。这件作品把焦点放在从亲身经历的角度发表自己动物伴侣相关帖子的个人，以文本来呈现人类鉴照在动物主观体验上的理解和激情。它意在提高对动物复杂生存状态以及人类与非人类之间关系的意识。《闪耀的蜂巢》制作期间的音景采用了"粒子合成"法，从录制的声音文件中生成实况，从而创造出不断变幻的恍惚的声音景观。文本和声音的群集从某个方面而言本身就是对群体意识的一种描摹。

Twitter 或 Sina Weibo 实时对话被编码，然后映射成动画的文本模式，令人想起昆虫的婚飞模式，可视地再现了共同经历与群体智慧。在装置内部，参观者还可以通过现场与电脑连接，或是从自己的移动装置上发帖到可视化文本中，从而实现互动。这件作品依据共同的探讨主题，将网上和现场通信的即时模式表现为一种集体效应和通信空间维度。

*glisten) HIVE is a project mapping real-time on-line conversations about animal consciousness into an audio-video interactive installation using computer algorithms that mimic social-insect swarming patterns. Animated text and sound generation depict Twitter or Sina Weibo contributions about animals. The project looks at individuals who are posting from the 1st person point of view of their animal companion. The project offers textual representations of human empathy and compassion reflecting on the subjective experience of animals. The project is intended to raise awareness about animals' complex states of being and human nonhuman relations. The soundscape during *glisten) HIVE uses "particle synthesis" methods, generated live from recorded sound files, to create an abstract ever-changing sonic landscape. The swarm of text and sound, in a sense, takes on a life of its own and depicts group consciousness.

Live Twitter or Sina Weibo contributions are coded and projected into moving text patterns reminiscent of swarming patterns of insects, visualizing shared experience and group intelligence. Within the installation, visitors are also encouraged to interact by posting entries to the visualization via an on-site computer and interface or from their own mobile devices. Here, the emergent pattern of on-line and on-site communities are represented as a collective effect and a spatial dimension of communication based on shared themes of discussion.

闪耀的蜂巢，2012
朱丽·安德烈耶娃、西蒙·奥威尔斯多（加拿大）
互动参与式音频、视频装置
© 朱丽·安德烈耶娃

*glisten) HIVE, 2012
Julie Andreyev, Simon Overstall (Canada)
Interactive, Participatory Audio, Video Installation
© Julie Andreyev

亲密，2012
卢森格德工作室 达安·卢森格德（荷兰）
交互技术
© 卢森格德工作室

Intimacy, 2012
"Daan Roosegaarde with Studio Roosegaarde" (Netherlands)
Interactive Technologies
© Studio Roosegaarde

《亲密》是一件时尚作品，探究私密与技术之间的关系。它的高技术服装采用电子箔片制成，当人们靠近或触碰到它，它就会变得透明起来。社交互动决定了服装的透明度，创造了一种亲密的体感游戏。

"Intimacy" is a fashion project exploring the relation between intimacy and technology. Its high-tech garments are made out of opaque smart e-foils that become increasingly transparent based on close and personal encounters with people. Social interactions determine the garments' level of transparency, creating a sensual play of disclosure.

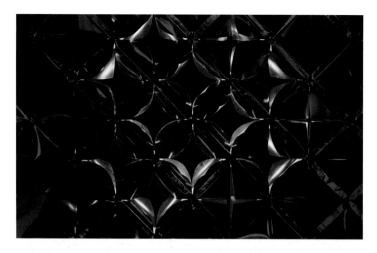

莲花, 2012
卢森格德工作室 达安·卢森格德（荷兰）
金属箔片
© 卢森格德工作室

Lotus Model, 2012
"Daan Roosegaarde with Studio Roosegaarde"(Netherlands)
Foil
© Studio Roosegaarde

《莲花》是一件有生命的艺术品,由智能金属箔片构成,它对人类行为作出响应,从而让花朵开合。当人们走过莲花,铝制薄片就以一种有机的方式自行绽开,创造出私密与公开之间透明的虚无空间。物理的隔墙变得无足轻重,而让位于空间与人类诗意的变换。

"Lotus" is a living artwork composed of smart foils that fold open in response to human behavior. Walking by "Lotus", aluminum foils unfold themselves in an organic way; generating transparent voids between private and public. Via "Lotus", physical walls are made immaterial, giving way to a poetic morphing of space and people.

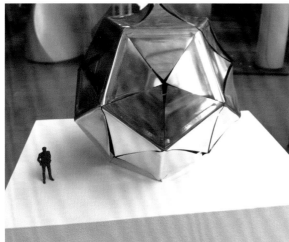

音乐织布机, 2005
伍韶劲(中国香港)
混合材料
© 伍韶劲

Musical Loom, 2005
Kingsley Ng (Hong Kong, China)
Mixed Material
© Kingsley Ng

这件作品把购自法国提花博物馆的一台有着 250 年历史的织布机转变成一套音像俱全的乐器。透过一面镜子，一块完整的屏幕被投射到纱线上，而红外摄像机和超声距离传感器则用来跟踪。参观者可以一一坐下来与机器互动。一根光条被吸附到参观者的手部，跟随着人手一起移动，让参观者能够生成机械的音景，也可以根据互动来随意地用音乐表情达意。参观者既可以在纱线上弹奏（如弹奏竖琴），也可以用手在空中弹起四声和弦。

The works transforms a 250 year-old loom (purchased from the Jacquard Museum of France) into a sound and image instrument. A single-screen is projected onto the threads through a mirror. An IR camera and ultrasonic distance sensors were used for the tracking. A participant can sit and interact with the machine one at a time. A light bar snaps to and follows the participant's hand movements. He or she then generates mechanical soundscapes, or malleable musical expressions based on their interaction. One can play on the threads (like a harp), and control 4-voice harmonies and volumes by the hands' positions in midair.

随着人类趋向于用三维方式来反映真实世界,这就不仅帮助我们定位数据,同时还让我们能够确定它们与环境的关系。现代数码世界中完全缺失这种品质,用户只能把自己的数据与文本或图像关联起来。空间和生动感方面丢失了巨量的信息,从而用户也降低了对数字数据的控制水平。道格拉斯构想了一种未来的三维接口,通过有形的控制器,用户就能用它来身临其境地体验三维数据。

As human beings tend to reflect the real world in a three dimensional manner, it helps us to not only locate data, but also its relation to the environment. In modern digital world, this quality is completely missing and users can only relate their data to text or images. Vast information is lost in terms of space and vividness, so as to decrease the level of control over digital data. Douglas has visualized a future of 3D interface with which users can experience immersive 3D data through tangible controllers.

混合现实系统——身临其境三维体验的明智解决方案，2011
王骁勇（道格拉斯）（加拿大）
木家具、电脑渲染系统、电脑视觉跟踪系统、背投屏幕
© Pleasantuser 设计技术公司

Mixed Reality System—a wise solution for immersive 3D experience, 2011
Wang Xiaoyong (Douglas)(Canadian)
Wood furniture, Computer Rendering System, Computer Vision Tracking System,
Rear-projection Screen
© Pleasantuser Design & Technology, Ltd.

该作品为美国Inwindow Outdoor公司为20世纪福克斯公司的电影《阿凡达》蓝光DVD发行仪式设计并制作的交互式户外宣传。该装置通过人脸识别技术捕捉到观众的面部影像，经过参数化的计算，实时地将观众的头像变形、拉伸、染色成为电影中的阿凡达形象。其中的一个技术亮点是在变形过程中，计算机会分析每个人不同的五官及面部特征，并影响每个局部的变形参数，使生成后的"阿凡达"头像还保留着该观众的特点。

The work is the interactive outdoors publicity designed and produced by America Inwindow Outdoor Company for the issuing ceremony of blue-ray DVD of the film "Avatar" of the 20th Century Fox. The device captures facial images of the audience with the face identification technology, and deforms, stretch and dye the portraits of the audience into the image of Avatar in the film on real-time basis through parameter calculation. One of the technical highlight is that the computer will analyze the different five sense organs and facial features of each person during morphing, which will influence the morphing parameters of each part, and retain the features of such audience in the portrait of "Avatar" generated thereby.

阿凡达变形站，2009
艺术指导、编程 / 师丹青（中国）
编程 / Mooshir Vahanvati（印度）
数字互动影像
© 美国20世纪福克斯，美国纽约 Inwindow Outdoor 公司

Avatar Morphing Station, 2009
Art directior, programming / Shi Danqing (China)
Programming / Mooshir Vahanvati (India)
Digital Interactive Image
© America 20th Century Fox, New York Inwindow Outdoor, America

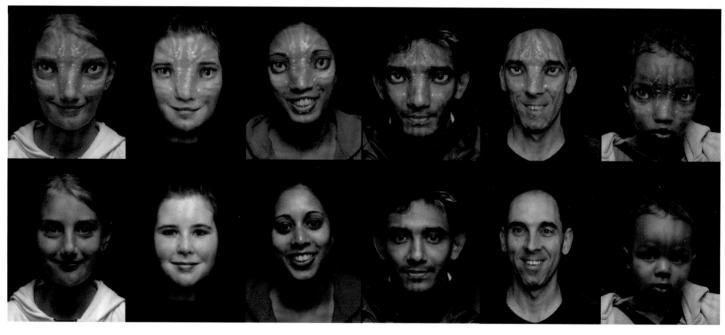

我的太极, 2012
祝卉（中国）
基于移动平台的交互软件应用
© 祝卉

My Tai Chi, 2012
Zhu Hui (China)
Based on Software Application and Video Display of Mobile Platform;
Interaction between the Works and Audience on the Site; Picture Display Board
© Zhu Hui

该作品将三维图形图像的采集、绘制、展示等技术运用在移动平台上，同时结合了混合现实与交互的技术，实现了三维动态教学的展示及娱乐互动。该作品获得APP STORE连续两周的推荐，并作为国家863课题的示范项目，同时入选ACM SIGGRAPH 2012。

2010年3月，项目负责人联系到了70岁的"中国陈氏太极第一人"的太极大师，并开始进行太极拳的数字化保护和应用研究。经过近两年的时间，项目负责人带领着课题组成员完成了陈氏太极一路、二路、十三式的数据化，以及移动网络平台上的新的呈现方式。

为了提高教学的便利，该作品除了能让用户360°实时观看角色动画，实时地动态换装，更换三维教练形象，更换教学场景外，还提供了动作对比的功能。一个即使不熟悉影片编辑软件的用户，也能轻松地调出自己的动作录像，实时地与三维角色的太极动作进行对照。同时，该作品增加了混合现实的元素，使得用户可以实时地将自己真实的脸更换在三维角色上。赶紧与朋友家人分享这样有趣的场景画面吧。

The works applies technology such as collection, drawing and demonstration etc. of 3-D pictures and images on the mobile platform, and realizes demonstration and recreational interaction with mix reality and interaction technology. The works has been recommended by APP STORE for two weeks successively, and has been enlisted in ACM SIGGRAPH 2012 as the pilot project of the state 863 project.

The person in charge of the project contacted the 70-year-old Tai Chi master reputed as "the primary master of Chinese Chen's Taiji" in March 2010, and carried out digital protection and application research on Tai Chi, and led the members of the project team to complete digitalization of one established series of skills and tricks, two established series of skills and tricks, and thirteen postures of Chen's Tai Chi, and new means of demonstration on the mobile network platform through nearly two years of work.

In addition to the panoramic real-time animation cartoons of the role play, real-time dynamic changes of clothes, and change of 3-D images of the coach and teaching scenes for the users, the works has also provided the function of comparing motions to facilitate convenience in teaching. Even a user unfamiliar with film editing software can easily find out its own motion video for real-time comparison with the Tai Chi motions of a 3-D role. Meanwhile, the works has added elements of mixed reality to enable users for real-time replacement of the face of the 3-D role with its own real face. Please lose no time to enjoy the interesting scenes with friends and family members.

《恒生态》是一个依托增强现实技术，通过手持移动设备摄像头，识别展览的主题LOGO，会在移动设备屏幕上展示一段以本展览LOGO主题延伸出来的三维现实增强动画。营造一个物理现实空间和虚拟数字世界并存的多维互动体验空间，内容涵盖数字世界、物理世界和精神世界等三个方面，追求信息、生态和智慧的平衡和谐发展。

作品在此基础上增加二维码参展作品识别的功能，通过将程序设想屏幕对准参展作品标牌的二维码，则可以显示该作品的图片作品信息、艺术家介绍等等扩展信息。

手机终端程序可以提供下载地址，支持ios4以上的iphone、ipad用户，同时展览现场为体验者提供一个ipad，在工作人员的协助下，将观者和作品内的现实增强数字动画LOGO进行合影留念，之后可以将合影发送到观众的邮箱。

"Permanent Ecology" is to display 3-D augmented reality animation extended from the LOGO theme of the exhibition on the screen of the mobile equipment by means of the camera of hand-held equipment used to identify the thematic LOGO of the exhibition based on augmented reality technology. The works builds up a multi-dimensional interactive space of experience with coexisting physical real space and virtual digital world, with its contents covering three aspects such as digital world, physical world and spiritual world, which pursues balanced and harmonious development of information, ecology and wisdom.

The works adds the function of identifying the works with 2-D codes at exhibition based on that, and focuses the screen on the 2-D codes of the signs of the works at exhibition through program to display the expanded information such as the information on the works, and introduction to the author etc..

The terminal program of the mobile phone can provide the following downloading address, and support users of iphone and ipad above iOS4. Meanwhile, the exhibition provides an ipad to the those who want to experience the works on the site. Photos are taken for the audience and the augmented reality digital animation LOGO in the works with the aid of the staffs, and then the photos will be sent to the mailboxes of the audience.

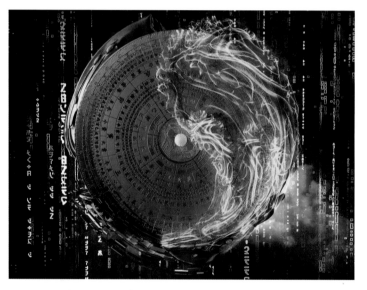

恒生态，2012
渠蒙（中国）
平面印刷品、ipad
© 渠蒙

Permanent Ecology, 2012
Qu Meng (China)
Plane Printed Works, ipad
© Qu Meng

东方的美学始终将宇宙和自然作为主体，而让我们谦逊得仰望宇宙的无限，从而使我们成为"道法自然"法则的守护人。

今天的中国每天都在不停地刷新速度，我们已习惯栖居在我们自己设计的钢筋水泥的高楼丛林中，而离"自然"越来越远。

回过头来，子曰的"智者乐水，仁者乐山；智者动，仁者静；智者乐，仁者寿"怡情传统自然的精神已经成为当代大部分人的梦想，不知这是一种悲还是一种喜？

现在，作者试图通过新媒体的水墨影像进行时空转换，让古意中的水墨山水影像成为宇宙信息的载体，随着观众的不断介入，让水墨的丝丝墨痕与悠扬的琴声渐起，而跟随着观众的身影在虚拟影像的空间里泛起层层的涟漪，让观众身体的舞动肆意惊动了水墨山水的视觉演绎，再让水墨意境随着观众的飘影进入而在虚拟景象中找寻自我的身影而忘怀于自然的化境中。

Oriental aesthetics always takes universe and nature as the mainstay, thus we have been made so humble as to look up at the infiniteness of the universe, and become the guards of the "natural rules".

Nowadays, China is setting a new speed constantly every day. And we have been accustomed to living in the high buildings made of reinforced concrete designed by ourselves, thus getting farther and farther away from "nature".

Looking back at the past, the spirit of soothing the mind with traditions and nature as shown in the words of Confucius "The wise enjoy water, the humane enjoy mountains. The wise are active, the humane are quiet. The wise are happy; the humane live long lives" has become a dream of most of the contemporary people. I wonder whether it is sorrow or joy.

At present, the author try to carry out conversion of time and space through the water-ink images of new medias, and convert the water-ink landscape images in the ancient sense into a carrier of universe information. The ink traces and the melodious musical sounds of piano will gradually come out with the constant involvement of the audience, which will produce boundless ripples in the space of virtual images with the shadows of the audience, and the rampant dancing of the audience's body will strike the visual evolution of water-ink landscape. Thus the water-ink scenes will seek own shadows in virtual images, and bring about natural realm with the entry of audience's flying shadow.

自然的另一种状态，2011
金江波（中国）
新媒体互动影像作品
© 金江波

Another Status of Nature, 2011
Jin Jiangbo (China)
New Media Interactive Video Work
© Jin Jiangbo

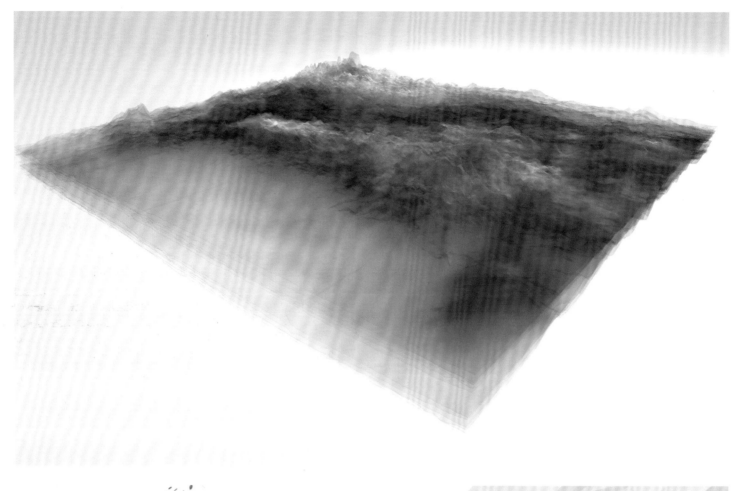
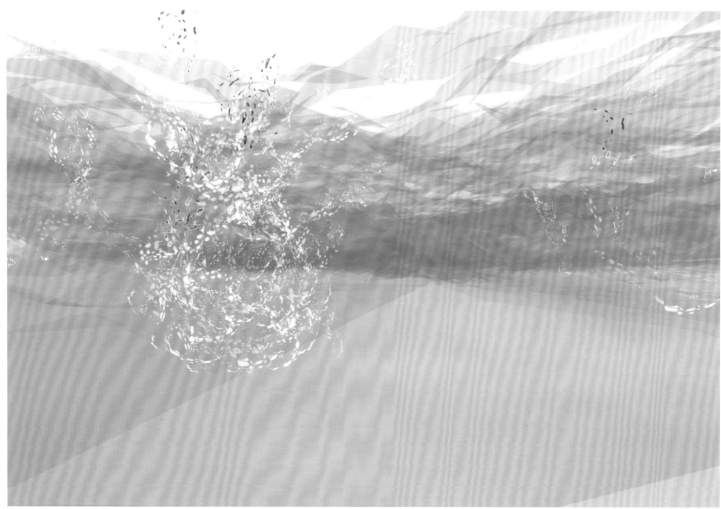

《纳米精华》旨在构建起一种亲身体验，以便从纳米的尺度上考察科学的世界与先验的世界。它是一套交互式视听装置，参观者可以通过自己的呼吸与配乐的画面进行交流。在这件作品中，呼吸与生命的开端有着牢固的概念性和隐喻性纽带。作品运用了以悬臂式原子力显微镜扫描人类表皮细胞从生到死所取得的数据，然后用它们来生成一系列分层的纳米拓扑图，用于大型的数据的投影，以呈现生命精髓的混合隐喻景象。凯文·拉斯沃思依据细胞自动机而开发出的一种算法生命形状在混合隐喻景象中浮现出来，形成一种生死之间的空间轮廓。参观者的呼吸以及从接口传感器中获得的水分影响着生命的形状，以生成并模拟让新生命形状得以生长的参数。

《纳米精华》中有声的拓扑图纹是从原子力显微镜以力谱模式记录的数据中分析而成的。人永生化表皮细胞原子的振动一开始是从体外扫描的，之后随着能量媒介一同注入血浆中，得出的数据在比较分析结果后又转换为声音文件。这件从听觉上呈现的作品包括了纳米尺度上发声的振动，同时也向听众提供一种可触的拓扑图感觉。

The "Nanoessence" project aims to construct a physical experience to examine scientific and metaphysical world at a nano level. "Nanoessence" is an interactive audio-visual installation where the viewer will interface with the visual and sonic presentation through his or her own breath. In the context of the project breath has a strong conceptual and metaphorical link to inception of life. Nanoessence uses data captured from an Atomic Force Microscopes cantilever scans of the living to dead HaCat skin cell. The data scans are used to generate a series of layered nano topographies for a large data projection display representing hybrid metaphorical landscape for the essence of life. An algorithmic life form developed by Kevin Raxworthy based on cellular automaton emerges within the hybrid metaphorical landscape that creates an envelope of space created between life and death. The life form is affected via the viewers breath and moisture gained from the interface sensors, to create and stimulate the parameters for the new life forms to grow.

The audible topographic texture of "Nanoessence" is created from analysis of data recorded from the AFM in force spectroscopy mode. The vibrations of the HaCat cell atoms are scanned initially in vitro and then with ether injected in to the serum. The resulting comparative analysis of the data is converted into sounds files. The installation presents the auditory work that consists of sonic vibrations that occur at the nano level and will also be presented to the audience as a haptic topographic sensation.

纳米精华, 2009
保罗·托马斯、凯文·拉斯沃思（澳大利亚）
代码、木材、速成原型模型
© 保罗·托马斯

Nanoessence, 2009
Paul Thomas, Kevin Raxworthy (Australia)
Code, Wood, Rapid Prototype Model
© Paul Thomas

太空纳米硅土气凝胶具有自然的天蓝色。瑞利散射是一种白光在尺寸小于光波长的粒子（往往为5—20纳米的粒子）上发生散射的光学现象。由于具有自然的蓝色及橘黄色（瑞利及米氏散射），太空纳米材料为本人所有作品的核心材料。本作品的安装高度为180—250厘米，因此观众无法触碰到星盘。但在遥控装置的帮助下，观众可以调整作品的亮度（LED及激光灯）。

The space nanomaterial silica aerogel has a natural blue color for the same reason the sky is blue: Rayleigh scattering, an optical phenomenon that results when white light scatters off of particles smaller than the wavelength of light, particles typically of the size 5-20nm.
The space nanomaterial is the center of all my artworks thanks to its natural colors blue and orange (Raylight and Mie scattering). This artwork has to be fixed in a height of more than 180cm and less than 250cm. Thus the viewer cannot touch the sky disc. Nevertheless with the help of a remote control (s) he can change the lighting in the works (LED and laser light).

禁止接触，2010
Ioannis MICHALOUDIS（希腊）
硅土气凝胶、纤维、木箱、LED及红色激光、遥控装置
© Ioannis MICHALOUDIS

Noli Me Tangere, 2010
Ioannis MICHALOUDIS (Greek)
Silica Aerogel, Tabric, Wood Box, LED and Red Laser Light, Remote Controls
© Ioannis MICHALOUDIS

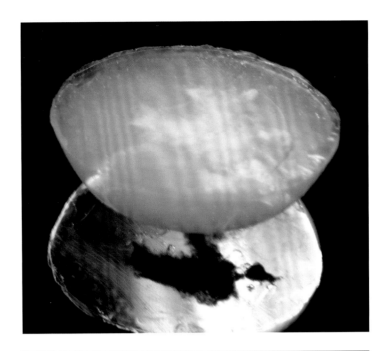
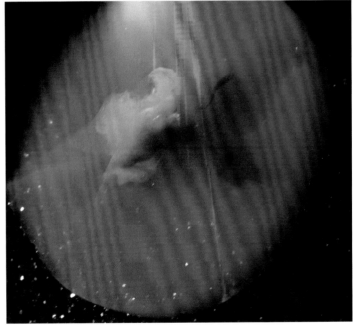
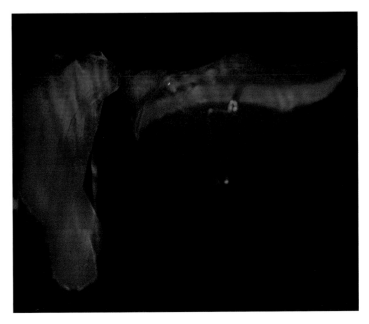

微软亚洲研究院研制的10亿级像素数字相机系统能够生成分辨率高达13亿像素的高质量图像,可用于文物和艺术品的高精度数字化采集。软硬件结合的配置,使数字相机系统能通过自身软件来加强硬件优势,大幅提高数字化拍摄的工作效率和性价比,同时也降低了拍摄难度。该系统可以精确自动地控制数字相机在较短时间内自动完成所有指定拍摄,大大减轻了拍摄者的工作量。软件的自动拼接功能可以将拍摄好的图片智能高效地自动拼接完成,而焦点合成技术可以高精度地捕捉几何立体细节,自动计算景深,分次拍摄多张同一场景但焦点不同的照片图像,然后将所有的图像图片合成,使得同一场景下的每一细微处都呈现焦点清晰的影像。此外,定制的俯仰平台满足了多角度文物和艺术品拍摄的需求。本作品通过把微观细节与宏观视角相结合,传达给观众这样的信息:这个世界既是简单的,也是复杂的。物质的形态在不同的尺度下有着其多样性,我们需要更加全面和完整地了解这个世界。

The billon-pixel digital camera system researched and produced by Microsoft Asia Research Institute can generate high-quality pictures of a definition rate of 1.3 billion pixels, which can be used for high-precision digital collection of cultural relics and art works. The configuration of combined software and hardware can strengthen the advantages in hardware through the own software of the digital camera system, and greatly improve the working efficiency and performance price ratio of high digital photography and lower the difficulties thereof at the same time. The system can accurately and automatically control the digital camera to complete all the appointed photo-taking automatically within a short time, and greatly reduce the workload of the photographer. The automatic splicing function of the software can complete automatic splicing of the pictures taken intelligently and efficiently, while the focus composite technology can capture the geometric solid details with a high precision, and automatically calculate the depth of field as well as take several pictures and images of the same scene with different focuses. Then it will synthesize all the pictures, and each detail under the same scene will display images with clear focus. In addition, the customized pitching platform meets the demands for taking photos of cultural relics and art works from multiple angles. The works combines the micro details and macro perspectives, and conveys such information as follows: the world is both simple and complex. The forms of substances are diversified under difference dimensions, and we need to have more comprehensive and complete understanding of the world.

十亿级像素数字图像,2012
徐迎庆、敖梦星、邓岩、陈刚、梁潇、马歆、黄郁驰(中国)
数字图像
© 清华大学、微软亚洲研究院

Billion-Pixel Digital Picture, 2012
Xu Yingqing, Ao Mengxing, Deng Yan, Chen Gang,
Liang Xiao, Ma Xin, Huang Yuchi (China)
Digital Image
© Tsinghua University, Microsoft Asia Research, Institute

声调是汉语言与生俱来的特质，我们世世代代口口声声地沿用下来习以为常，作为母语的无意识的使用错似乎淡化了其本身的感知，反倒是日益庞大的中文学习者时常感佩汉语中惊人而难以掌控的声调。

《把玩声调》是这样一个能唤起人们对汉语特质的声调可视化交互装置。它将实时语音中的汉字声调进行捕捉，首尾相接，进而生成平、上、去、入的连续波形几何图形。图形之间的堆叠、交叉、错落等关系构成了各种生动活泼的游戏。参与者可以根据声调构造自己需要的声调图形，亦可即兴发挥。

The tones is the natural born characteristic of Chinese. Generation by generation, word by word, we inherit the functionality of them. Using Chinese as the mother tongue, Chinese people hardly sense the wave of tones that generate when speaking certain words. In contrast, increasing number of Chinese learners often time feel the function of tones is the most difficult part of Chinese.

"Play With Tones" is such a vocal interactive installation that trying to evoke and rediscover the beautiful nature of Chinese tones. It basically captures the tones while the participants speak, and then put the tones together so it become an undetermined wave that stretch on the screen. The variation and the pattern of tones form a context of various games. The participants can either build the certain wave according to the pattern, or make improvisation while playing.

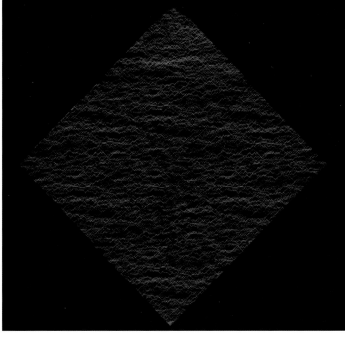

把玩声调，2011
郭耀、蒋程宇（中国）
多媒体
© 郭耀、蒋程宇

Play with Tones, 2011
Guo Yao, Jiang Chengyu (China)
Multi-media
© Guo Yao, Jiang Chengyu

三只大气球任意飘荡在展厅内，观众发出一点动静就会激活它们，无论是出自人们的互动，还是空气的流动，气球的任何移动都会激发出声音，好像来自空气中的下雨声，这些气球就像天然的内置扬声器，把雨声放大出来。

Three giant black sound balloons floats arbitrary around the gallery. Any movement activates atmospheric sounds of rain. The Floating Spheres are designed to remain buoyant in the gallery space. Any movement of the balloons spheres, by people interacting or air movement in the gallery, activates sounds, as such; the balloons act as a natural boombox, amplifying the sound.

气球雨，2012
蒂娜·贝赫（丹麦）
混合材料
© 蒂娜·贝赫

Rain Balloons, 2012
Tine Bech (Danish)
Mixed Material
© Tine Bech

作品是"利用光学原理与物理错视原理,以运动的发光载体,塑造神秘的信息时光隧道空间,呈现具有深远空灵的心理效应"。通过 LED 光源与光效结合,以新媒体互动艺术的表现形式,突出表达现代数字技术与图像艺术的相互交融,内容与形式的相互配合,让观众沉浸在数码时代多种信息刺激的幻想与现实的模糊空间中,以此形成本作品的设计特点。
造型形式:作品为一个直径 1.2 米的圆形体(直径 1.2 米 × 厚度 0.35 米),悬空挂置造型。互动形式:自动感应为互动形式;采用红外线控制的互动技术,随着观众挥手的动作,控制时光空间的色彩与运动,炫目多彩,使观众感受到"穿越时空"那互动效果的惊奇、互动的精彩。

The works builds up mysterious tunnel space in information times and displays profound, free and natural psychological effects with a moving luminous carrier by making use of optical principles and the illusion principles in Physics. The combination of LED optical source and optical effects highlights the mutual integration of modern digital technology and video art with the means of expression of new media interactive art; Mutual cooperation between contents and forms soaks the audience in the illusion of multiple information irritation in the digital times, and dim space in reality to form the design features of the works.
Model Form: The works is a sphere with a diameter of 1.2 meters (diameter 1.2 meter * thickness 0.35 meter), suspended model. Means of interaction: automatic induction; infra-controlled interaction techniques; The hand-shaking movements of the audience controls the colors and movements of the space of time and space, which is brilliant and enables to the audience to experience the magic and interactive effects of "passing through time and space".

时光机器,2012
周春源(中国)、Chen Cui(加拿大)、谢昭一、广州美术学院陈小清新媒介工作室
金属、有机玻璃、LED 光源
© 周春源

Time Machine, 2012
Zhou Chunyuan (China), Chen Cui (Canada), Xie Zhaoyi,
Chen Xiaoqing New Media Studio of Guangzhou Academy of Fine Arts
Metal, Organic Glass, LED Light Source
© Zhou Chunyuan

展翅飞翔是人类最古老的梦想之一。费斯托公司从银鸥获取灵感,创造了智能飞鸟（Smart Bird）,成功破解鸟类飞翔的秘密。智能飞鸟可以自如地起飞、翱翔并降落,而无需借助额外的驱动装置。这是因为采用了主动关节式扭转驱动单元与复杂的控制系统组合,从而获得了前所未有的飞行效率。通过分析智能飞鸟的飞行参数,费斯托获得了更多的知识优化其产品和解决方案的能效性。

With "Smart Bird", Festo has succeeded in deciphering the flight of bird—one of the oldest dreams of humankind. Inspired by herring gull, Smart Bird can start, fly and land autonomously—with no additional drive mechanism. This is made possible by an active articulated torsional drive unit in combination with a complex control system attains an unprecedented level of efficiency in flight operation. By analyzing SmatBird's fly characteristics, Festo has acquired additional knowledge to optimize its products and solutions in an even more energy efficient manner.

智能飞鸟, 2011
费斯托公司（德国）
混合材料
© 费斯托公司

Smart Bird, 2011
Festo AG & Co. KG (Germany)
Mixed Material
© Festo AG & Co. KG

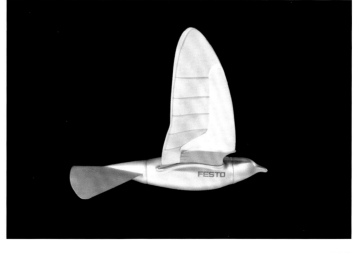

费斯托仿生学习网络

费斯托成立于1925年,是全球领先的自动化技术供应商,并提供先进的技术培训和专业再教育课程。公司总部位于德国埃斯林根,是一家独立的家族企业。

自1990年代起,费斯托就致力于仿生技术研究,向大自然学习提高能效的经验。在费斯托看来,生物演变就是机体结构为适应环境而进行优化的过程。2006年,集合了知名大学、研究所和企业的仿生学习网络成立,旨在将自然界的能效原理转化成自动化技术和制造工艺,并开发出能效型的仿生机电一体化产品。无论是银鸥,还是大象鼻子,抑或是壁虎爪子上的吸盘都成了费斯托灵感的来源。

如今,利用仿生学习网络开发并优化自动化技术已成为费斯托研发工作的重要平台之一。

Festo Bionic Learning Network
Established in 1925, Festo is a worldwide leader in the field of automation technology and the global market leader for technical training and continuing vocational education. The independent family-owned company is headquartered in Esslingen, Germany.

Festo has been working intensively on the topic of bionics since early 1990s, believing nothing is as efficient as nature: biological evolution can be described as an optimisation strategy for the adaptation of organisms to their environment. In 2006, the Bionic Learning Network, an association of renowned universities, institutes and development companies, was launched, with the aim to transfer natural efficiency principles to automation technologies and manufacturing processes and to develop energy-efficient bio-mechatronic products. Herring gull, elephant trunk or the pads of a gecko can all serve as the inspiration.

Today, with the aid of bionics the Bionic Learning Network serves as a platform in addition to the R&D activities of Festo to develop and optimize new technologies in automation.

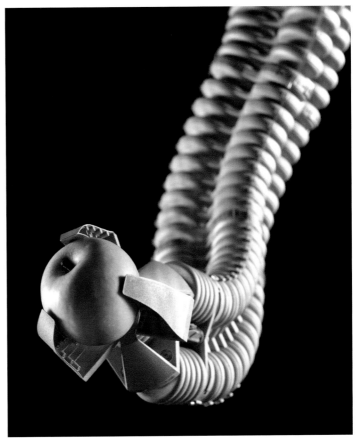

《爱面子》是一件网络表演装置，适合在热闹的公共场合中比如博物馆和广场上表演。在展览期间，将一部智能电话固定在站台上。

这件作品展示了在网络时代人们遭遇到信任、可视性与私密性方面的问题，关注人们情感上的反应和社会的反应。《爱面子》是智能约会的一种可视的、诗意的仪式，它邀请参与者触摸、抚爱自己的脸面，用这种方式来构成一种在线联网的身份，从而与北京的亲朋好友或陌生人交往接触。

"Saving Face" is a hosted performance & installation. During exhibition the smart phone is placed in a fixed booth. "Saving Face" takes place at dynamic public spaces like museum, square.

"Saving Face" shows our emotional and social encounter with trust, visibility and privacy in our contemporary smart cities. "Saving Face" is a visual and poetic ritual for smart meeting. It invites participants to touch and caress their own face; and in this way to compose an online networked identity to connect with family, friends and strangers in Beijing.

爱面子，2012
凯伦·兰塞尔、赫尔门·马特（荷兰）
混合材料
© 凯伦·兰塞尔、赫尔门·马特

Saving Face, 2012
Karen Lancel, Hermen Maat (Netherlands)
Mixed Material
© Karen Lancel, Hermen Maat

Sifteo 的物联网智能平台试图打造一种新颖的交互式游戏体验。你可以用手来移动、摇撼、旋转或变换排列积木的方式，使积木之间形成相互传感。

作品试图融汇两种重要的游戏传统——将象棋、多米诺以及拼板这类典型的游戏模式与丰富的交互式娱乐技术相结合，从而创造出令人兴奋、富有挑战性的互动感受。

每一个智能积木都配有一张全彩 LCD 屏、多种运动传感器和可充电电池，这一切都包含在 1.5 英寸大小的固体方块中，供任何年龄层次的人娱乐。这套设备需通过无线接口与近邻的计算机相连，充电一次可运行 4 小时。

Sifteo's Intelligent Play™ Platform is a fresh, tactile take on interactive play. Use your hands to move, shake, flip, rotate and neighbor your Sifteo cubes—you'll be entertained and inspired by Sifteo's unique gameplay.

At Sifteo, it is bringing together two great play traditions, combining classic play patterns from games like chess, dominoes and jigsaw puzzles with the richness of interactive game technology for an experience that is exciting, challenging and fun.

Each Sifteo cube packs a clickable, full color LCD display, a variety of motion sensors and a rechargeable battery into a sturdy 1.5 inch block—perfect for hands of all ages to grab and play with. Sifteo cubes connect wirelessly to a nearby computer via a compact USB radio link, and can hold up to 4 hours of play on a single charge.

智能积木，2011
Sifteo 公司（美国）
物联网交互设备
© Sifteo Inc.

Sifteo Cubes, 2011
Sifteo Inc.(USA.)
Wireless Interactive Cubes
© Sifteo Inc.

《沉睡的花》是一个由透明的金字塔锥体和立台组成的交互装置。观众可以看到金字塔内有一株发着光晕的花朵，花瓣微微摆动，如同在沉睡中呼吸。当敲击立台唤醒花朵时，花瓣会随着敲击声舒展开来，光晕也随之发生变化。

《沉睡的花》利用光学的原理形成裸眼三维影像，并结合艺术的表现为观众带来一种超越真实世界的视觉感受。同时，自然而简单的互动在作品和观众之间架起了一座桥梁，为观众提供了良好的体验。

"Sleeping Flower" is an interactive installation that consists of a transparent pyramid and a box for knocking to interact with the contents shown in the pyramid. Audience will find a 3D flower glowing with naked eyes in the pyramid, the flower petals waving, like breathing in a sleep. When audience knock the box to wake up the flower, the petals will dance with the knocking sound, the color of glow will also change.

The installation brings art and science together to get real 3D views, invoking a different visual perception of the world. A natural yet simple interaction bridges the gap between the works and audience, providing a nice experience of interacting with the artwork.

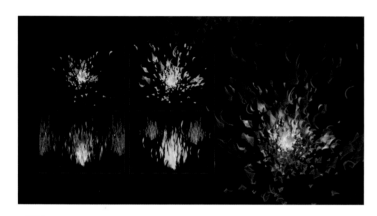

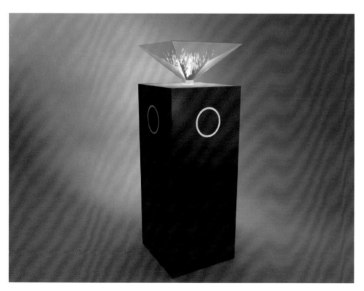

沉睡的花，2012
设计 / 吴琼（中国）
技术支持 / 邹冠上、佟馨、易鑫、张靖宁、许默涵
综合材料
© 吴琼

Sleeping Flower, 2012
Design/ Wu Qiong (China)
Technical support/ Wu Guanshang, Tong Xin, Yi Xin, Zhang Jingning, Xu Mohan
Mixed Material
© Wu Qiong

风筝文化是我国重要的非物质文化遗产。由于生活节奏的加快、活动场地的局限，放风筝的人越来越少了。而老的手工艺人的流逝，风筝制作技艺的失传，以及外国高科技风筝的引进等原因，使风筝这一重要的非物质文化遗产正在面临濒临灭绝的危机，急需得到人们的重视和保护。本项目通过利用微软的 KINECT 技术，我们可以将风筝文化进行技术创新，将濒临失传的中国传统风筝文化与 Kinect 技术结合，实现风筝的室内放飞，并实现风筝这一中国濒临失传的传统文化的传播和非物质文化遗产的数字化保护。

Kite culture is an important intangible cultural legacies of China. Due to the faster life rhythms, and limitations of site, there are fewer and fewer of kite flier. Moreover, the lapse of old craftsman, loss of kite manufacturing skills, and introduction of foreign high-tech kites etc. have brought kite, an important intangible cultural legacy, to the verge of extinction. Thus kite is in urgent need of attention and protection. With Microsoft KINECT technology in the project, we can carry out technical innovation of kite culture, and combine Chinese conventional kite culture on the verge of loss and KINECT technology to realize indoors kite flying, and spread of kite as conventional culture on the verge of loss, and the digital protection of non-intangible cultural legacy.

放风筝的人，2011
赵月、张爽爽、于阳、魏一明（中国）
混合材料
© 赵月、张爽爽、于阳、魏一明

The Kite Runner, 2011
Zhao Yue, Zhang Shuangshuang, Yu Yang, Wei Yiming (China)
Mixed Material
© Zhao Yue, Zhang Shuangshuang, Yu Yang, Wei Yiming

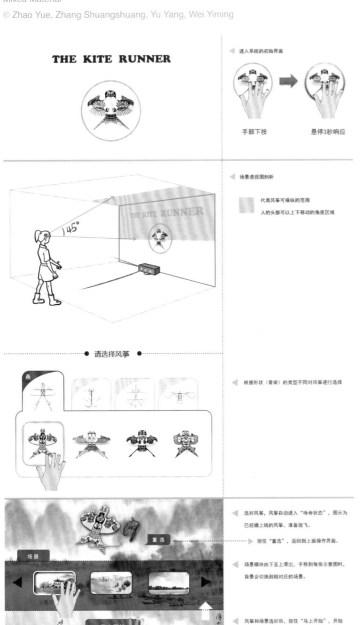

079

城市图谱, 2012
米兰理工大学（意大利）
图像、视频
© 米兰理工大学

Urbanscope, 2012
Milan Politecnico (Italy)
Images, Video
© Milan Politecnico

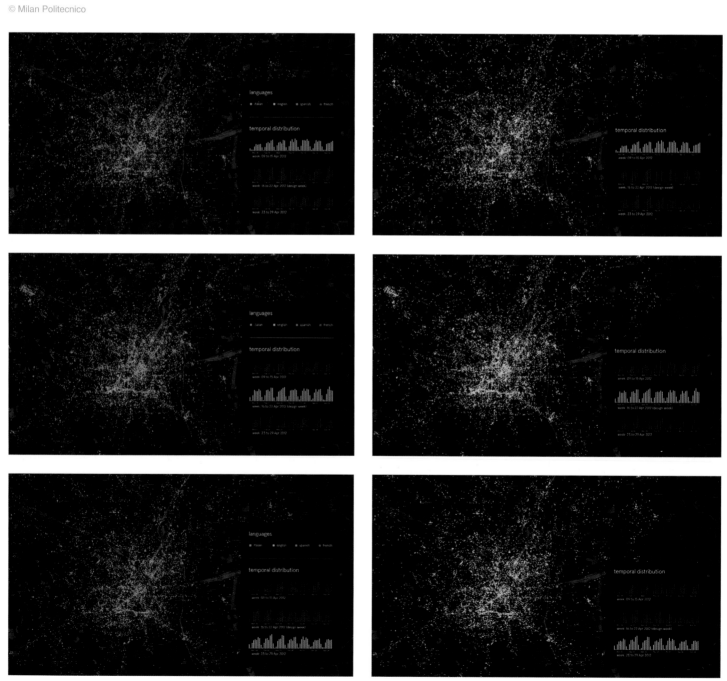

《城市图谱》是一项持续的研究作品,它扫描多层复合的城市景观,突出它多声、多中心的多重性,并聚焦于生活、居住或穿行于城市地区的人们留下的数字踪迹(比如地理定位的社群媒体活动、用户生成内容、手机通话等),以探寻他们对所在城市的观点、向往和记忆。在《城市图谱》中,采用了新的方法和技术,使这些数据在不同的情形下具有各种艺术和文化价值。这些新方法能够收集、分析和呈现城市规模的实时数据以及定量指标,使我们能够判断城市中如何定义人员和地域,并通过人们留下的数字踪迹来使这些判断可视化。

城市图谱聚集了学术界和专业界的智慧,创造新的知识和工具,旨在设计、测试并应用收集、分析和呈现城市尺度实时数据的新方法与新技术,启发我们为城市尺度的未解疑问寻觅可能的答案。

"Urbanscope" is an ongoing research that focuses on digital traces left by the people who live, inhabit and cross urban areas (such as geo-localized social media activity, user generated content, mobile phone activity…) in order to investigate their perceptions, visions, memories of the city. Within "Urbanscope" we're designing testing and deploying new methods and technology to make those data worth in different situations: new methods able collect, analyze and represent real-time data as well of qualitative indicators at the urban scale, to determine how people and places are defined and made visible from the digital traces they leave.

"Urbanscope" is a transdisciplinary think tank that involves academic and professional partners in the creation of new knowledge and tools able to enlighten potential answers to unsolved questions at the urban scale. "Urbanscope" aims at designing, testing and deploying new methods and technologies to collect, analyze and represent real-time data at the urban scale. "Urbanscope" is an urban periscope that scans the composite, stratified, reactive urban landscape to highlight its plurivocal and polycentric multiplicity, to make its polyphonic images visible.

城市中严重的汽车尾气污染，不仅夺走了昔日的蓝天，威胁到人体的健康，也使人们对未来的生活感到沮丧。《会呼吸的灯-1》正是基于这一问题而提出的解决方案。该设计源于植物的"呼吸作用"和"光合作用"。它以太阳能和风能为动力，一方面将城市中的"尾气"吸入"体内"，经"空气净化装置"转变成清洁空气后排出"体外"；另一方面通过顶端的LED照明装置为城市提供高效的照明。《会呼吸的灯-2》是一个延展方案，其特点在于它的扇叶具有显像功能。扇叶中心的"摄像头"可以捕捉到附近人的动态，然后通过体内的微处理，将相应的图像动态地显示在扇叶上，实现人和物的"交互感应"。

The serious pollution from urban auto exhaust has not only snatched the blue sky in the past, threatened human health, but also thrown people into depression toward future life. "Breathing Lapm-1" is the solution program put forth based on such issue. The design originates from the "respiratory function" and "photosynthesis" of plants. It is powered by solar energy and wind energy, and inhales urban "exhaust" inside on the one hand, and discharges clean air outside through conversion by the "air purifying device". On the other hand, it provides highly-efficient lamination for the urban areas with the LED lighting device at the top. "Breathing Lamp-2" is an extension program, which is characteristic of the image displaying function of its fans. The "video camera" in the center of the fans can capture the movements of people nearby, and dynamically display the corresponding images on the fans to realize "interactive induction" between people and objects.

会呼吸的灯，2012
邱松（中国）
综合材料
© 邱松

Breathing Lamp, 2012
Qiu Song (China)
Composite Materials
© Qiu Song

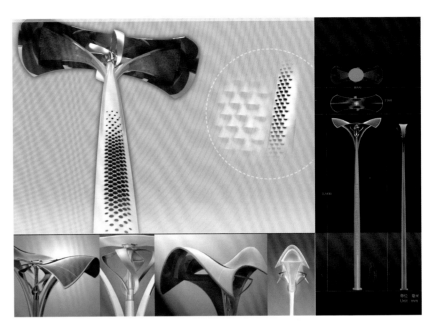

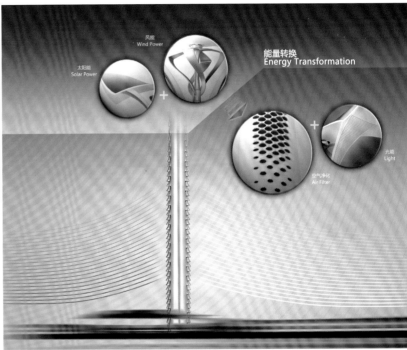

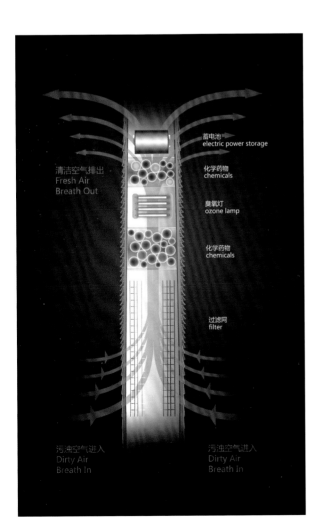

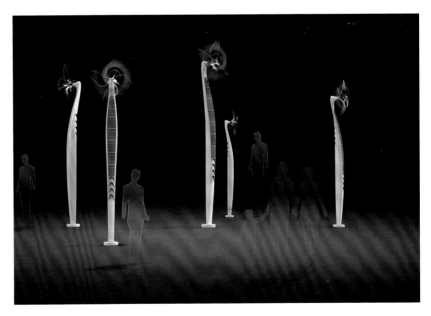

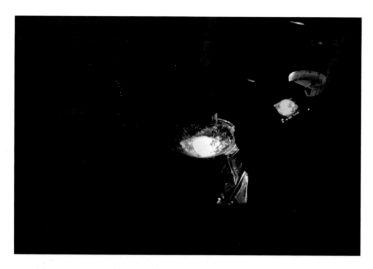

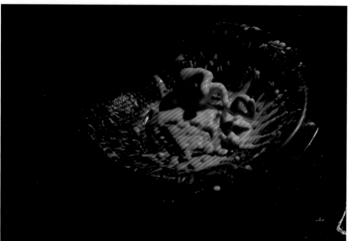

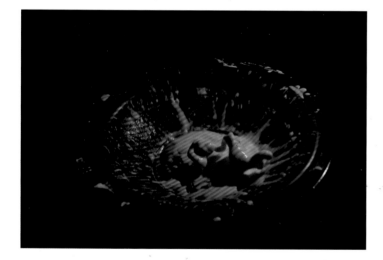

这是一组体现时间中某一时刻的雕塑作品。它的灵感来自对声音和振动如何影响物质、如何创造自身内在形式的研究,整个作品产生于有机材料的处理过程。

"音流学"一词源于希腊语"波浪",它研究可视的声响与振动,后者属于模态现象中的一个细分领域。正如德勒兹在他的《斯宾诺莎》这部著作中所指出的那样,"形式来源于物质自身内在的东西,物质并非外来形式的惰性受体"。这组雕塑作品便采用了不同的形式起源的框架,在这种框架中,形式并非由外界强加给物质的。

《音流嬉戏》这个作品聚焦于视听效果的双重呈现,以及材料的形成过程。

A series of sculptures representing moments in time obtained from the research of how sounds and vibration affect matter and can create form immanent to itself feeding from material processes with organic life like behaviour.

"Cymatics" (from Greek "wave") is the study of visible sound and vibration, a subset of modal phenomena. As Deleuze shows in his works on *Spinoza*, resources involved in the genesis of form are immanent to matter itself instead of seeing "matter as an inert receptacle for forms that come from the outside". Working in this frame of an alternative model of the genesis of form, one in which form is not imposed on matter from the outside.

"Cymatics Disport" lies between audiovisual performance and form finding processes through material properties.

音流嬉戏,2012
豪尔赫·拉米雷斯(墨西哥)
非牛顿流体(玉米淀粉+水)、扩音器、放大器、CD 播放器、定制软设备、以不同的频率和声速来播放、投影、沙雕
© 豪尔赫·拉米雷斯

Cymatics Disport.2012
Jorge Ramirez (Mexican)
Non-newtonian Fluid (Cornstach+Water), Speaker+Amplifier+Cdplayer, Custom Made Software to Run Different Frequencies and Subsonics, Projection, Sand Sculptures
© Jorge Ramirez

"瓦伦西亚理工大学(UPV)当代艺术藏品"
手袋上印有从瓦尔特本雅明《机械复制时代的艺术作品》中摘录的一段话,"复制技术使物体从传统领域中脱离,从而成了某种独有存在的大批副本……。" 这些文字被打散开来,分布在每一只手袋上。如此一来,一段哲言所文饰的表面使物体有可能成为超出其本身功能的"内容容器"。
《内容提包》是专为瓦伦西亚理工大学制作的、聚焦于当代艺术的收藏品。它放在校园内,与人们的日常融为一体,让生活的形式与文化的内涵结成复杂的交互关系,同时也具备跨学科的特性。
它让我们能够以某种移动容器(手袋)来展现这所大学艺术收藏的一部分(即:女性艺术家作品),这些容器用绘画、雕版印刷等各种图像来"装扮",从而让人想象我们所拥有的物品具备什么样的新内容。

-An Art collection " the Contemporary Art Collection of Polytechnic University of Valencia (UPV)"
-Its reproduction " the images of the Works in this Collection"
-An object " a handbag, on whose surface these works will be printed "

A sentence from the text by Walter Benjamin "The Work of Art in the Age of Mechanical Reproduction" is printed on the handbag. The writing, fragmented, flows through each one of the handbag. In this way, the epidermis generated by a philosophical text makes it possible for the object to become a "container of contents" going beyond its real function.
"Containers of Contents" was a project tailored for the UPV focusing on its contemporary art collection. The collection is housed within the University, where Artworks share the space with the sensitivity and daily tasks of its people. This allows a permanent and complex interrelation between the forms of life and the forms of culture.
With this analogy the project is linked to the University environment, precisely for being a "Containers of Contents"; for the interrelation Art-Knowledge, and for the interdisciplinary character it features.
A proposal that allows us to show a part of the art collection of the University (made by women artists), by means of some mobile containers (the handbags) that "are dressed" with images from paintings, engraving and sculptures. And they perform it to provide the objects we imagine or own with a new content and meaning.

内容提包, 2012
伊莎贝尔·特里斯坦(西班牙)
手袋、塑料艺术复制品、影印照片
© 瓦伦西亚理工大学

Containers of Contents, 2012
ISABEL TRISTÁN (SPAIN)
Leather, Plastic with Art Reproduction, Photocopy
© UNIVERSIDAD POLITÉCNICA DE VALENCIA

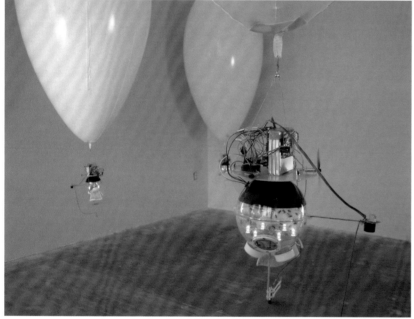

飞翔的软气艇，2010
戴维·博文（美国）
塑料、铝、电子产品、家蝇
© 戴维·博文

Fly Blimps, 2010
David Bowen (USA)
Plastic, Aluminum, Electronics, Houseflies
© David Bowen

《飞翔的软气艇》探讨的是互动、反应以及生殖过程中产生的美学问题，因为这些过程关联到自然与机械系统之间的交叉。作者制作了一些装置并创造出某种特殊情境，使这些装置处于运动变化之中，根据它们对所占时空的感知以及与这些时空的互动而产生出绘画、运动、文章、音响和物体。作者将机械结构和特定的情境相结合，再融入"机械"生命，最终形成本作品。难以预料的事件以及这些装置产生的独有结果使得新系统运动起来的概念令人痴迷。这一过程可以被视作是特意的，也可以是一种荒唐的审美数据捕捉方法。

《飞翔的软气艇》这件作品包括一组充注氦气的自主式气艇，它们的运动由一小群家蝇来控制。这些家蝇实际上就是这组装置的头脑，决定着这些气艇如何与空间以及其他装置互动，如何作出响应。每一只气艇内安装有一个隔间，里面生活着多达50只的家蝇。隔间中放有食物和水，还能采光，从而让这些苍蝇存活、繁衍。隔间中还安装有传感器，侦测苍蝇运动所产生的光线变化模式，它们实时地将这些信息传送给一只装载在气艇上的微控制器，后者启动与气艇推进器相连的电机，它们依据苍蝇的行动来引导气艇的方向。漂浮游弋的气艇象征着相互分隔，但也相互交汇的社会。苍蝇在自己自给自足的世界中存续，它们创造出对群体行为的一种引申、夸张的表达。

The works is concerned with aesthetics that result from interactive, reactive and generative processes as they relate to intersections between natural and mechanical systems. I produce devices and situations that are set in motion to create drawings, movements, compositions, sounds and objects based on their perception and interaction with the space and time they occupy. The works is the result of a collaboration between the mechanisms and situations I employ and the life of the "machine" itself. Unpredictable occurrences and unique outcomes that result from these devices fuel the fascination to set new systems in motion. This process can be seen as an elaborate and even absurd method of capturing aesthetic data.

The "Fly Blimps" consists of a series of autonomous helium filled blimps whose movements are controlled by small collectives of houseflies. The flies are essentially the brain of each of the devices, determining how they interact and respond to the space as well as the other devices. Up to 50 houseflies live within the chambers attached to each blimp unit. These chambers contain food, water and allow the light needed to keep the flies alive and flourishing. The chambers also contain sensors that detect the changing light patterns produced by the movements of the flies. In real-time, the sensors send this information to an on-board micro-controller. This controller activates the motors connected to the propellers that direct the devices based on the actions of the flies. The floating, wandering blimps are separate but intersecting community vehicles. The flies exist in their own self-contained and self-sustaining worlds, collectively creating an amplified and exaggerated expression of group behavior.

建筑中的气候响应性往往被理解为大量技术及电子感应、激励及调节装置。与叠加在惰性材料上的高科技设备相比,自然界体现了一种完全不同的非技术策略:在多个生物系统中,响应能力实际上已经深植于材料本身之中。

该项目采用为材料系统进行实际编程的类似设计策略。该策略不要求采用任何机械或电子控制措施,也不要求提供外部能力,而材料计算对环境进行反馈的方式。

气候敏感性形态学装置漂浮在一个完全透明的玻璃盒子中。在这个盒子中,气候对应于有关巴黎相对湿度的加速数据库。这个盒子的作用并非是从蓬皮杜中心地区(可被证明为世界上最稳定的气候区之一)分离出来,而是提供一个通向外界的虚拟连接,并展示了我们通过系统自身的无声变化难以感知的微妙的湿度水平变化。这些周期性的变化伴随着由一个访客蒸汽散发数据集内的阀值过渡引发的自发性气候事件。

所产生的自主性及被动式表面激励汇合了环境及空间体验。通过精巧及无声的气候敏感性建筑形态学装置加强了对微妙的局部不断变化的环境状况的感知。不断变化的表面状况实际上体现了在材料本身内部进行感知、激励及应对的能力。

Climate-responsiveness in architecture is typically conceived as a technical function enabled by myriad mechanical and electronic sensing, actuating and regulating devices. In contrast to this superimposition of high-tech equipment on otherwise inert material, nature suggests a fundamentally different, no-tech strategy: In many biological systems the responsive capacity is quite literally ingrained in the material itself.

The project employs similar design strategies of physically programming a material system that neither requires any kind of mechanical or electronic control, nor the supply of external energy. Here material computes form in feedback with the environment.

The meteorosensitive morphology floats in a fully transparent glass case. Within the case the climate corresponds to an accelerated database of the relative humidity in Paris. In this way, the case functions less as a separation from the interior space of the Centre Pompidou, arguably one of the most stable climate zones in the world, but rather provides a virtual connection to the outside, showing the subtle variations in humidity levels that we hardly ever consciously perceive through the system's silent movement. These cyclic changes are interspersed with spontaneous climate events triggered by threshold transitions within a second data set of visitor vapour emission.

The resultant autonomous, passive actuation of the surface provides for a unique convergence of environmental and spatial experience. The perception of the delicate locally varied and ever changing environmental dynamics is intensified through the subtle and silent movement of the meteorosensitive architectural morphology. The changing surface literally embodies the capacity to sense, actuate and react, all within the material itself.

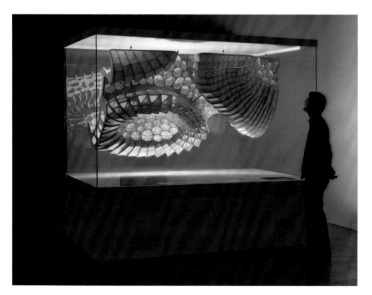

湿度计——气候敏感形态学装置,2012
阿希姆门格斯、斯蒂芬·赖克特(德国)
枫木贴面、桦木胶合板
© 阿希姆门格斯、斯蒂芬·赖克特
图片:蓬皮杜

HygroScope--Meteorosensitive Morphology, 2012
Achim Menges, Steffen Reichert (German)
Laminated Maple Veneer, Birch Plywood
© Achim Menges, Steffen Reichert
Photo: Pompidou

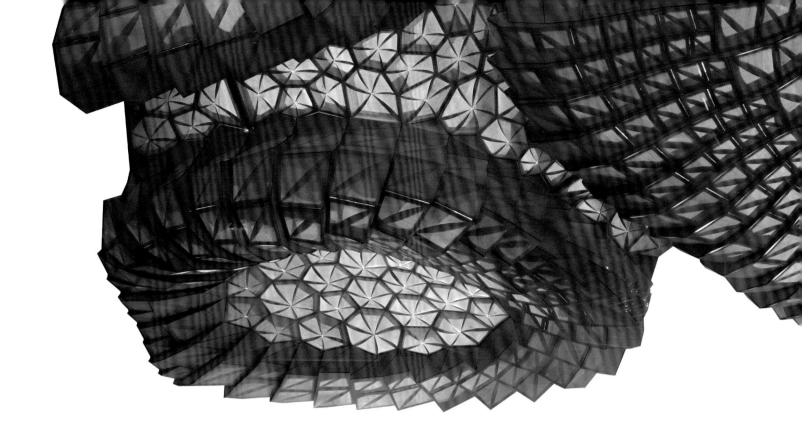
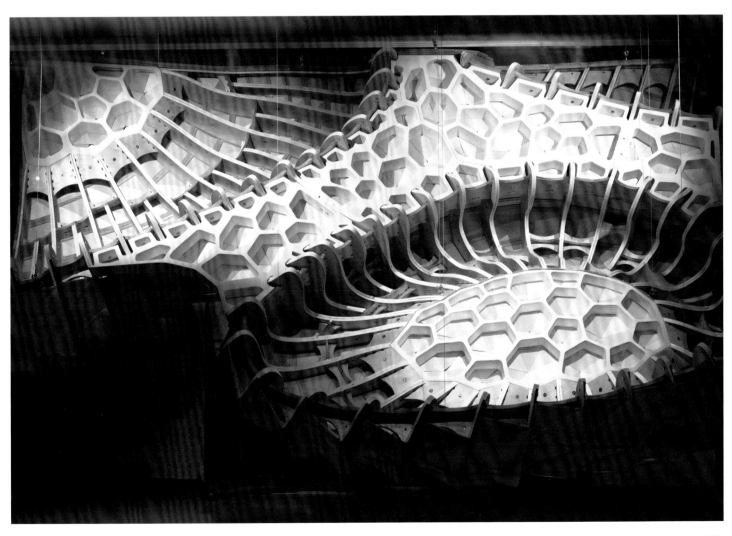

该设计通过微型水净化系统，实现卫浴空间的水资源可持续利用。结合湿地净水技术，将生态环境引入卫浴空间，带给用户绿色生态化的洗浴体验。

为迎接未来老龄化社会和信息化社会的挑战，运用通用设计方法，将卫浴空间设计成为既能满足社会弱势群体的需求，又能满足其他大部分用户的洗浴体验的综合化产品/服务系统。

运用云技术和人工智能技术，实现用户在洗漱过程中对于唾液、体重、脂肪等生理指标的健康检测，并将数据实时与医院共享，实现对用户健康的全方位关怀；运用生物检测技术，实现用户在如厕过程中对肝功等多项健康生理指标的化验检测，并实时提供健康报告和健身计划；实现在卫生间的健身功能，在生态卫生间良好的环境下，结合前述的健康报告和健身计划，用户可以利用创新设计的健身器材进行有计划的锻炼。

与手机软件相连接，进行个性化设置，同时测试马桶的功能与维修周期。通过卫浴空间内的产品介质，整合数字化技术实现洗浴过程中的娱乐功能（如利用镜子实现视频和音频娱乐），丰富用户的洗浴体验。综合运用新技术为用户提供全新的洗浴体验，如水温可视化设计、个性化定制服务等。

九牧未来整体卫浴概念设计，2012
赵超、杨冬江、范寅良、陆轶辰（中国）
不锈钢
© 清华大学、九牧集团

Future Conceptual Design of Integrated Bathroom Units of JOMOO, 2012
Zhao Chao, Yang Dongjiang, Fan Yinliang, Lu Yichen (China)
Stainless Steel
© Tsinghua University, JOMOO Group

The design realizes sustainable utilization of water resources in the bathroom space by means of micro water purification system. It introduces ecological environment into the bathroom space with the wetland water purification technology, bring about green and ecological bathing experience to users.

To meet the challenges brought about by the future aging society and informational society, the works applies general design methods to design the bathroom space into a comprehensive product/service system that can meet the demands of the social disadvantageous groups, and the bathing experience of most of the other users. In addition, it applies cloud technology and artificial intelligent technology to realize health check of physiological indexes such as saliva, body weight and fat etc. during washing, and can realize real-time data sharing with hospitals for comprehensive care for the health of the users. Moreover, it applies biological testing technology to realize chemical tests of several health and physiological indexes such as liver function etc. when relieving themselves, and can provide real-time health report and body building plan. And it can realize body-building function in the bathroom, and the users can make use of the innovatively designed body-building facilities for planned exercise in the desirable ecological bathroom according to the aforesaid health report and body building plan.

It is connected with mobile phone software for individualized configuration, and the function and repair cycle of the commode. It can realize recreational function during bathing (for example, realizing video and audio recreation with mirror) by means of the product medium and integrated digital technology in the bathroom space to enrich the bathing experience of the users. The comprehensive application of new techniques provides brand-new bathing experience to the users. For example, visible design of water temperature, and individualized customized service etc..

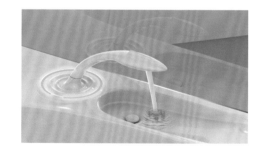
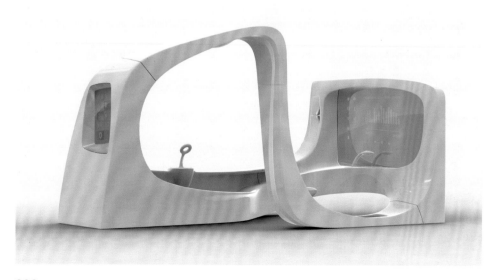
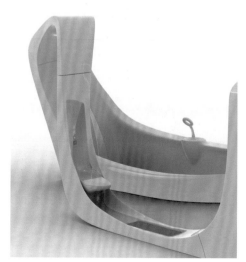

作品以生态树脂材料为基础依托，借用科学性实验手段将代表着自然生命的有机材料、纹理、天然元素等视觉符号任意组合及打散，在基因柱中尽情展现，不同颜色、面貌的基因柱通过不同的排列组合方式产生千变万化的生命万象图。借助观者的互动行为而产生千变万化的视觉效果，使人们在此种行为中体会生命与自然间玄妙和谐的关系，在艺术中感受生命的奥妙，在科学中探讨无限的可能。

Based on the ecological resin materials, the works carries out random combination and diffusion of visual symbols such as organic materials, texture, and natural elements etc. representing natural life by means of scientific experimental methods, and fully display them in the gene pillar. The gene pillars of different colors and features produce the figure of ever changing life phenomena through different means of arrangement and combination. Various visual effects are produced through the interactive behaviors of the audience to enable people to experience the subtle and harmonious relations between life and nature in such behavior, and the mystery of life in art and infinite possibility in science.

生命 印象，2012
刘强（中国）
天然植物、合成树脂
© 刘强

Life Impression, 2012
Liu Qiang (China)
Natural Plant, Composite Resin
© Liu Qiang

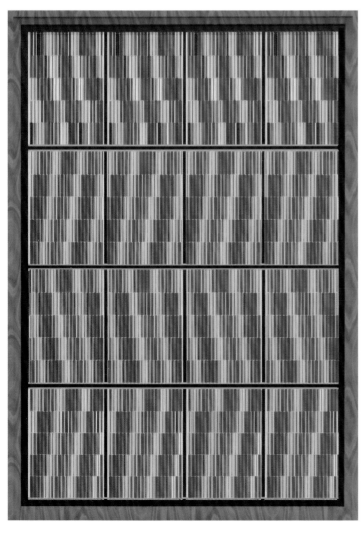

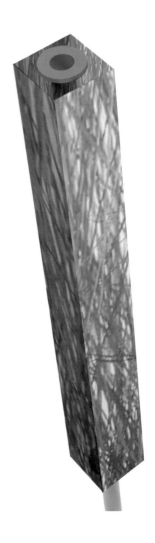

KIA RHYME 轿跑车是一款中高端双门四座油电混动轿跑车。从设计定位上，该设计用户群是一线城市中的中等收入家庭，他们也都有了新的消费观念，不再偏向功能主义，他们的需求正在呈多样化趋势。而汽车对他们来说不再单指一个快捷安全的代步工具，这也使得汽车逐渐从"档次身份符号化"逐渐向"性格品位符号化"转变。从材质上来说，该设计使用软性材质作为车辆表面材料，摆脱金属材质给人的高光的、冰冷的感觉，也将减少高硬度车体对人身的伤害，减小机动车辆给人潜意识中的威胁性与侵略性。造型语言上，旨在研究全平滑曲线给人以温和无害而又不失速度感的形态特征，并在车辆造型意向上继承了KIA-Kee的家族脸谱以保证"RHYME"的血统纯正。

'KIA RHYME" race car is a medium and high-end double-door four-seat oil-electricity hybrid race car. The targeted users of the car are the medium-income families in the frontline cities, for they have developed new consumption concepts, and will not deviate to functionalism. Thus their demands are being diversified. Automobiles are not merely a fast and safe riding tool for them, thus automobiles have gradually changed from "symbol of level and identity" to "symbol of character and tastes". As for the materials, soft materials are used in the design as the surface materials of the vehicle, and have got rid of the excessively bright and cold feelings of metallic materials, and will also reduce damage to human body by the vehicle of a high hardness, and thereat and invasion left over in human subconscious by the motor vehicles. As for the modeling language, it is intended to study the gentle, harmless and rapid morphological characteristics left by complete smooth curves on people, and the family face of KIA-Kee has been inherited in the modeling intent of the vehicle to guarantee the pure blood of "RHYME".

KIA RHYME，2012
李白（中国）
ABS
© 李白

KIA RHYME，2012
Li Bai (China)
ABS
© Li Bai

Peugeot-Eco 是针对 2020 年中国城市年轻用户设计的一款概念跑车，受到公共交通发展的影响，2020 年的年轻人对于汽车的需求将更加注重造型及其所展现的气质，同时绿色环保将成为新的时尚潮流，因此，我主要在这个设计的过程中对未来 2020 的跑车造型趋势进行了深入的探究，同时动力来源采用纤维素乙醇技术提取出的醇燃料，符合未来年轻人的崇尚绿色环保的理念。

Peugeot-Eco is a conceptual race car designed for the young urban users of China in 2020. Due to influences of the development of public traffic, the youths in 2020 will pay more attention to the model and the quality demonstrated by the automobiles. Meanwhile, environmental protection will become a new fashion. Thus I mainly carry out in-depth exploration into the trends in the model of future race cars in 2020 during design, and the driving force is the alcohol fuel extracted with cellulose alcohol technology, which complies with the concept of advocating environmental protection among future youths.

Peugeot-Eco Concept 2020，2012
黄豪（中国）
ABS、透明压克力
© 黄豪

Peugeot-Eco Concept 2020，2012
Huang Hao (China)
ABS, Transparent Chocolate
© Huang Hao

母亲与孩子，始于 1998 年，重新制作于 2012 年
阿克罗伊德 - 哈维工作室（英国）
滞绿草、黏土、黄麻
©Ackroyd & Harvey

Mother and Child, Negative 1998, regrown 2012
Ackroyd & Harvey (UK)
Staygreen grass, Clay, Jute
©Ackroyd & Harvey

阿克罗伊德与哈维一直在用垂直表面上从草籽生长起来的青草来作为鲜活的摄影介质（光合作用成像）。经过对光线的控制，新生草叶通过叶绿素的产生而具备了非凡的表达能力，既能摹写下简单的阴影，也能描绘出复杂的摄影图像。从某种意义上来说，阿克罗伊德和哈维将在感光胶片上制作照片的摄影艺术转换成苗草的光敏感性，以草叶浅黄深碧的晕色重现在黑白相纸上显现出的色调范围。每一片新发草叶都产生出与照射它的光线程度相当的叶绿素分子，而它产生出的绿色浓度将取决于它受到的光照强度。

Ackroyd & Harvey have been using grass grown from seed on vertical surfaces as a living photographic medium. In controlled lighting(photosynthesis photograph), the emergent blade has an extraordinary capacity to record either simple shadows or complex photographic images through the production of chlorophyll. In a sense, Ackroyd & Harvey have adapted the photographic art of producing pictures on a sensitive film to the light sensitivity of young grass and the equivalent tonal range developed in a black and white photographic paper is created within the grass in shades of yellow and green. Each germinating blade of grass produces a concentration of chlorophyll molecules that relates to the amount of projected light available to it and the strength of green produced is according to the intensity of light received.

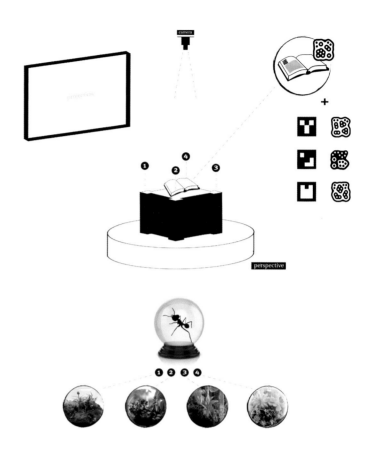

与蚂蚁等昆虫游戏，2012
会申（厄瓜多尔）
混合材料
© 会申

Playing With Ants & Other Insects, 2012
Kuai Shen (Ecuador)
Mixed Material
© Kuai Shen

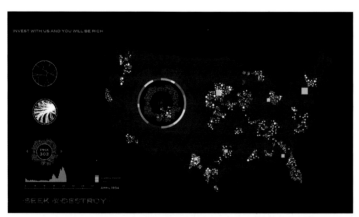

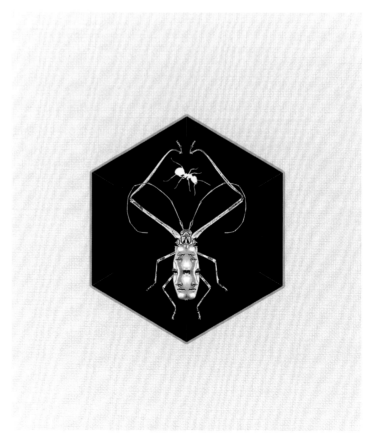

《与蚂蚁等昆虫游戏》是一项跨学科研究，旨在探究蚂蚁等昆虫潜在社会性。作品借鉴了 Roger Caillois 和 Jakob von Uexküll 所采用的行为学和后人文主义进行物种间研究的方式，这一作品特别侧重于探索"模仿"、"角色扮演"在人类游戏文化中所起的作用，以及人类游戏与一些昆虫神奇有趣的伪装和变形行为之间的联系。这些昆虫利用模仿，变换不同的外形、颜色和行为，在环境中隐形，以适应特定的生态系统和关系。此外，自维多利亚时代起昆虫对人类社会和文化一直有重大影响，这表现为：它们激发小说创作和超现实主义文学艺术家的灵感；成为供生物分类鉴定之用的昆虫收藏品，或成为反面乌托邦和极端性的怪异象征。

该项研究还旨在揭示生物游戏中的生物系统与人类设计的新兴游戏系统之间的联系。研究生物活体、人类和人造智能体在行为浮现与决策过程方面存在的类似性，对探究成就、竞争和协作具有非常重要的意义。蚂蚁的通信和社会行为作为一种游戏形式，与不断变化的人类视听技术环境有很强的相关性。蚂蚁和其他昆虫参与后人类生态学的构建和转变，这种后人类生态学的未来发展依赖于物种之间的沟通以及人造物品与生物有机体的互惠共生。

"Playing with Ants & Other Insects" is an interdisciplinary research that explores the potential aspect of social play in ants and other insects, building on ethological and post-humanistic approaches that arise predominantly from the interspecies discourses of Roger Caillois and Jakob von Uexküll. This thesis will specially focus on the role of mimicry in the human culture of play, mainly in role-playing games, and their association with that strange and fascinating behavior of some insects that camouflage and transform using mimicry, assuming different forms, colors and acting to disappear in the environment in order to adapt to specific ecosystems of relationships. Furthermore, since the Victorian era insects have played a major role influencing human society and culture as inspiration for fiction writing and surrealism, as entomological collectibles for taxonomy identification, or as a monstrous fascination towards dystopia and polarity.

The interest of this research is also to reveal the connections between biological systems in the game of life and those game systems designed by humans by means of emergent game play. It is of essence, to investigate the existing parallels in the emergence of behavior and decision-making process of living organisms, humans and artificial agents when it comes to achievement, competition and collaboration. Special attention is focused on ants' communication and social behavior as a form of emergent game play that strongly relates to the constantly changing audiovisual technologic milieu of humans. Ants and other insects participate in the construction and mutation of a post human ecology, an ecology whose future lies in the communication between species and the mutualism between human artifacts and biological organisms.

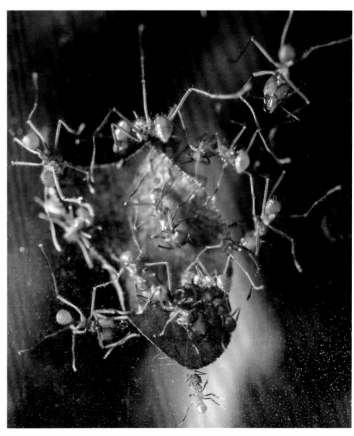

2012年美国航空航天管理局科学家预计会出现下一次超级太阳风暴——"电磁海啸",类似1859年9月的情形,当时世界各地都能看到激光,新生的电报技术遭到自然射电的干扰,而那时人类还没有发明无线电。2001年扬-彼得·E·R·索恩塔格实验室开始依据威廉·杜德斯1900年发明的"歌唱的弧光"装置来研制接口。这就是SonArc的发端,它是一种采用交替变换的审美格式来进行的项目过程,以寻觅电的真髓以及驯养闪电的可能性。

1989年,美国的"亚特兰提斯"号太空梭在执行STS-34飞行任务中第一次观察到了持续3微秒的巨大红色闪光。阿拉斯加大学的戴维斯·森特曼把这种低光度TV摄像头拍下的常压等离子体称作"小妖精",它跟北极光一样,只不过速度要快得多,仅仅一闪而过。"天空是硬物质在地球上方形成的巨大拱顶,其中有一个大窟窿,人的精魂便从这里通向真正的天国。……这就是极光……有时候伴随着极光的'噼啪'的尖啸便是这些精魂想要与地球上的人相互交流时发出的声音。"人类学家E·W·霍克斯在他1916年的著作《拉布拉多的爱斯基摩人》中这样描述因纽特人的"天国区域"。因纽特人能听到北极光的声音,亚历山大·冯·洪堡在18世纪也提到过这种声音。洪堡是提出北极光与磁场之间关系的滥觞者。光可以被描述为一种电磁波。我们所称的射电也可以用同样的方式来描述,只不过它的波是看不见的,也就是不可见光。我的实验室通过互联网与三座太空观察站连接,从而开发了一套软件,它同时从不同位置记录尖啸和"噼啪"声的数据流,这其实就是电离层的天电,等于极低频电波,然后将它们以声波的形式投射到感知空间中——这就是"小妖精"和北极光变换的电磁回声的声音形状。我在19世纪60年代的"盖斯勒管"(真空玻璃管)基础上重新设计,通过这些管子创造了通过真实北极光来触发的人工北极光。

1882年,科学家塞利姆莱姆斯特罗姆在拉普兰的一座山上制造了第一台、也是唯一一台能够生成人工诱发北极光的机器。两年前我得到了莱姆斯特罗姆1886、1887年间科学研究著述的三卷原稿,这是十分稀贵的资料。如今我拥有早先全部的方案,可以重新开始这项实验——创造与沃尔特·德·玛利亚的"闪电力场"相类似的艺术"极光场"。

等离子体是一种导电气体,它包含游离的载荷子(离子和电子),能够产生各种发光现象,其中最著名的便是太阳风击中地球大气上层而产生的北极光。19世纪时,研究者开始在实验室中用这些游离的能量来进行实验,想要在玻璃管中制作出人工极光,这批研究者中就包括芬兰科学家塞利姆·莱姆斯特罗姆,两年前扬-彼得·E·R·索恩塔格得以从拍卖会上购得他的论著和数据表。如今美国航空航天局以及美国海军都在用这些电波和电流进行研究。对于索恩塔格而言,等离子体就是中心物质,不仅因为它是泰斯拉能量的优美例证,还因为它可以被看作是传播媒介及其在20世纪光芒四射的显示的基础。

这两套装置都用当今的手段再现了19世纪的实验室情境。在一张工作台上有一只荧光灯管,工作台下面安装了一只所谓极低频电波的接收器。通过高频控制,这些电波——没有导体——便让荧光灯管闪烁微光。另一个房间中的工作台上也出现了类似的情况。墙壁上贴着尼古拉·泰斯拉这位19到20世纪之交重要的发明家和研究者的专利和设计图,与之相并的是最近研制出的手机交流电电源的图解,从而强调了泰斯拉研究工作的可行性。

In 2012 NASA scientists expect the next solar super storm, an "electromagnetic tsunami", like in September 1859 when auroras could be seen all over the world and the new technology of electric telegraphy was disturbed by natural radio before our radio had been invented. In 2001 Jan-Peter E.R. Sonntag and his lab started to develope interfaces based on Williams Duddels "singing arc" from 1900 to code pure plasma. It had been the start of the SonArc project: a project cycle in alternating aesthetic formats in search of the essence of electricity as well as the possibility of domesticating lightening.

For the first time in 1989 a red big flash of three milliseconds had been observed from the Space-Shuttle STS-34. Dr. Davis Sentman from the University of Alasca named this atmospherical plasma seen by a low-light-level TV camera: Sprite. Sprites are the same like nothern lights but much faster like a flash. "The sky is a great dome of hard material arched over the earth. There is a hole in it through which the spirits pass to the true heavens. (…) This is the light of the aurora. (…) The whistling crackling noise which sometimes accompanies the aurora is the voices of these spirits trying to communicate with the people of the earth." —the anthropologist E. W. Hawkes reported about the "heavently regions" of the Inuits in his book *The Labrador Eskimo* in 1916. The inuits can hear the aurora borealis what also Alexander von Humboldt in the 18th century mentioned. Humboldt had been the first one who had mentioned a relation between the nothern lights and the magnetic field. Light can be described as an electromagnetic wave. What we call radio can be described in the same way but the waves are unvisible = invisible light. Connected via internet with 3 space-observation-labs my lab has developed a software to stream simultanously from different positions whistlers and cracklers = sferics from the ionosphere = ELF (Extremely Low Frequency) waves - projected as soundwaves into the perception space – the sonic figures of transformed electromagnetic echos of sprites and nothern lights (light figures). Based on Geissler tubes (evacuated glass tubes) in the sixties of the 19th century I have redesigned those tubes to create through them atificial northern lights triggered by real aurora beorealis.

The scientist Selim Lemström built the first and only machines on a mountain in Lapland which created artificial induced aurora borealis in 1882. Two years ago I aquired the rare original three volumes of Lemström's scientific studies from 1886/1887. Now I have all the original plans to re-inact this experiment--an aurora-field in art analog to the lightning-field of Walter de Marias."

Plasma is a electrically conductive gas consisting of free charge carriers (ions and electrons) able to produce various light phenomena. One of the best known phenomena is aurora borealis (the northern lights) which is caused by solar wind hitting the upper strata of the earth's atmosphere. During the 19th century, researchers started to experiment with these free energies in laboratories and tried to produce artificial auroras in glass tubes. One of these researchers was Selim Lemström from Finland, the volumes and tables of whom Jan-Peter E.R. Sonntag could purchase by auction two years ago. Today, NASA and the US Navy are doing research with these waves and currents. To Sonntag, plasma simply is the central matter; not only because it is a beautiful example for Tesla energy, but because is can be considered the basis of the mass media and their luminous displays in the 20th century.

Both installations reproduce a 19th century laboratory situation using today's means. On the one table, there is a fluorescent tube. Beneath the table top a receiver of so-called extreme-low-frequency waves has been attached. By use of a high frequency control, these waves – without a conductor – set the fluorescent tube gleaming. Similar things happen on the table in the other room. On the walls, patents and design drawing by Nicola Tesla, the important inventor and researcher at the turn of the 19th to the 20th century are juxtaposed to recently developed schemata for the alternative power supply for mobile phones, thus emphasising the actuality of Tesla's research.

等离子体，2001—2012
扬 - 彼得·E·R·索恩塔格（德国）
混合材料
© 扬 - 彼得·E·R·索恩塔格

PLASMA, 2001-2012
Jan-Peter E.R. Sonntag (German)
Mixed Material
© Jan-Peter E.R. Sonntag

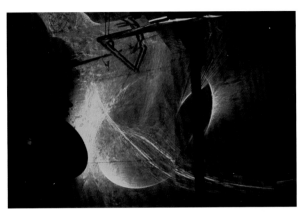
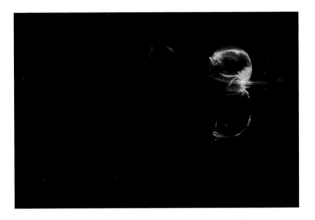
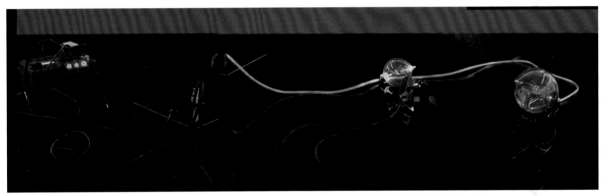

海滩怪兽/仿生兽（系列作品），1990年至今
奇奥·詹森（荷兰）
塑料管
© 奇奥·詹森

Strandbeests, since 1990
Theo Jansen (Netherlands)
Plastics Tubing
© Theo Jansen

《海滩怪兽》系列装置作品是有关艺术与自然的自主性对话。作者泰奥杨森毕业于代尔夫特理工大学物理系,其后却转而学习绘画,他相信"艺术与科学的界限只存在于我们心中"。从1990年开始,杨森便致力于"动感雕塑"的创作,至今已坚持22年。经过不懈的努力,他用塑料管和塑胶瓶的组合模式印证了雕塑可以拥有自己的生命甚至自由地迈步行走。也由此创造了一群巨大、多足的海滩怪兽。这些人造"动物"完全由黄色中空的塑料管捆扎塑造而成,它们行动的能量并不来自于食物,而是来自海滩的风能。通过简单的物理效应,其足底末端的塑料管(触角),可以感受沙滩的湿度,当行走到太过湿润或干燥的沙地上时,它们会朝相反的方向移动,这实际上保证了海滩怪兽的足迹不会偏离海岸线。塑胶瓶制作的物理感应器还可以在暴风雨将至的时候使它们自动打桩,将自己固定在沙地上。

按照杨森的设想,未来的海滩怪兽能够进化出自己的神经系统、肌肉和各类感应"器官",这一切都与生态环境交相辉映、浑然天成。这种自行在海滩上生存进化的动感雕塑实际上是艺术生命在自然生态中自我适应的行为映射,是超越了人类主观行动意志的自主化的艺术存在与生态表达。在这里我们看到装置艺术借助简单的物理媒介创造出新的生命存在,并为人类思考家园、思考生态、思考美丽的大自然提供灵感。

It's not every day that you run across an entirely new strain of life, which is exactly what Dutch kinetic sculptor Theo Jansen has created. His Strandbeests are wondrous wind-powered automatons that exhibit an incredibly lifelike dexterity as they cascade in flowing waves down seaside sands. The elegantly articulated creatures are constructed using genetic algorithms and are constantly evolving to better suit their environment.

The creatures are also able to store air pressure and use it to drive them in the absence of wind: "Self-propelling beach animals like Animaris Percipiere have a stomach. This consists of recycled plastic bottles containing air that can be pumped up to a high pressure by the wind." Theo's more sophisticated creations are able to detect once they have entered the water and walk away from it, and one species will even anchor itself to the earth if it senses a storm approaching. Watching a herd of them crossing a windswept beach would certainly be a sight to behold.

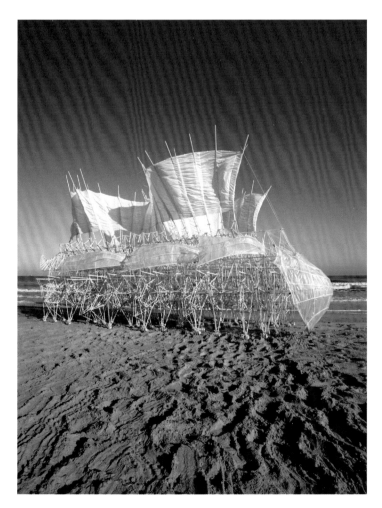

海峡动力水轮机，2012
安东尼·雷亚莱（美国）
综合材料
© 创意设计学院（美国）

Strait Power Turbine, 2012
Anthony Reale (USA)
Mural Relief Made from MDF
© College for Creative Studies (USA)

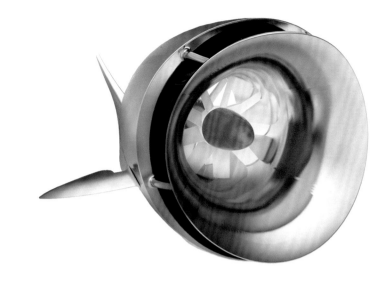

海峡动力水轮机通过艺术、设计和技术而使人类重返生态平衡系统，它以可扩展、可持续的环保方式来运用低速水流的动力。这个设计受到姥鲨（鲨鱼的一种）天然形状的启发，采用工业时代的技术来制造外形、交互作用与自然相谐的水轮机。

"Strait Power Turbine" is man's reintroduction into the ecology through art, design and technology. Strait Power is a turbine that harnesses power from low velocity currents in a scalable, sustainable, and environmental way. Strait Power is inspired by the natural form of the Basking shark and uses industrial-age technology to produce a turbine whose form and interaction are harmonious in the environment.

天堂的轮廓（可持续照明），2012
厄休拉·达姆（德国）
蚊虫、扩声器、水族箱
© 厄休拉·达姆

The Outline of Paradise (Sustainable Luminosity), 2012
Damm, Ursula (German)
Midges, Loudspeakers, Aquarium
© Damm, Ursula

如果不用人工照明，而用像萤火虫一样光闪闪的群舞蠓虫来制成广告牌，我们的城市会是怎样一番光景？《天堂的轮廓》探究了技术科学的前景与能力，并从中研制出一套如前所述的装置。

我们为这套装置培养了一群不咬人的蚊虫，让它们以特定的方式而飞舞，以使它们群飞的形状构成广告的内容。这些昆虫根据我们在培养和声音方面的资料来经过基因改造，能够在黑暗中发光。这种初步培养的结果将代相传承，从而保持它们群舞的形状。

我们怎么能教会昆虫认识字母？怎么能教会蚊虫认识字母表？由于蚊虫对声音敏感，我们就使用了一套实时的声音空间化系统来教它们。截至目前我们还只能培养出构成简单的 LED 字形的蚊虫群。

天然蚊虫的群舞形状是循环的圆形。群飞的蚊虫中包括聚集起来求爱的成年雄虫，它们通过拍打翅膀的声音而编排队形。我们的系统便利用了蚊虫对声音的这种敏感，以合成的拍翅声来组织它们。蚊虫只对它们自己种属拍翅的声音敏感，这种声音通常是特定频率 ±50 赫兹。要教会蚊虫识别字母表，我们便在这一频率范围中将字母编码。通过扬声器的空间定位，蚊虫便记住了这些声音的频率，学会以特定方式来对复音作出反应，从而在群飞时也学会了跟字母相关的集体行为。

What would our cities look like if advertising messages are produced not from artificial lighting but from swarming midges, glowing like fireflies? The "outline of paradise" explores the promises and capabilities of technoscience and developes and installation out of this narratives.

For the installation we train non-biting midges (chironomidae) to fly in a way that their swarm takes the shape of advertisement messages. The insects are genetically modified to glow in the dark and alter their genetic make-up according to the training and sound input we provide. This initial training will be inherited over generations and keeps the swarm in shape.

How can letters be tought to insects? How can we teach the alphabet to midges? As chironommidae are sensitive to sound, we use a real-time sound spatialisation system to teach the midges. Until now we are only able to produce clouds of midges forming a simple LED font.

Natural midges (chironomidae) form swarms with the shape of a circulating sphere. The swarms consists of male adults congregating for courtship. They are organized through the sound of the wingbeats of the male midges. Our system uses the sensitivity of chironomidae for sound and organize them with synthetic wing beat sound. Midges are sensitive to sounds within the range of the wing beat of their species. This sounds are normally ± 50Hz around the specific frequence. To teach midges the alphabet, letters are coded with nine different sounds withing this range. Through the spatial placement of the loudspeakers midges learn to react in a certain manner to polyphonic tones by memorizing sound frequencies and the letter-related collective behaviour of their swarm.

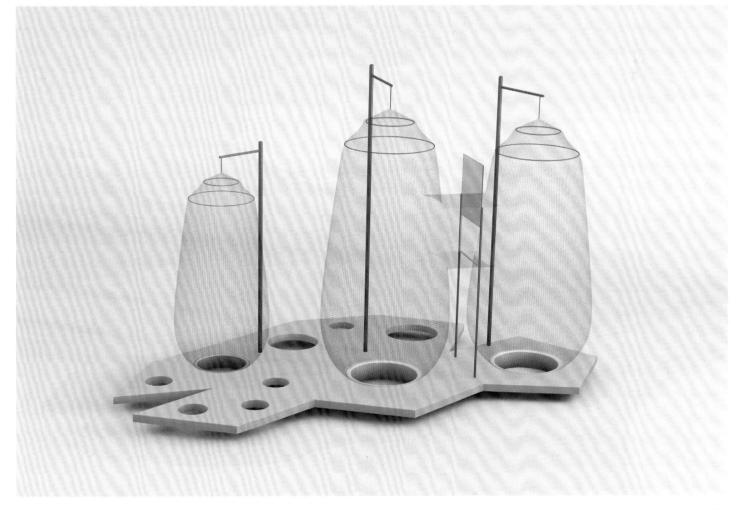

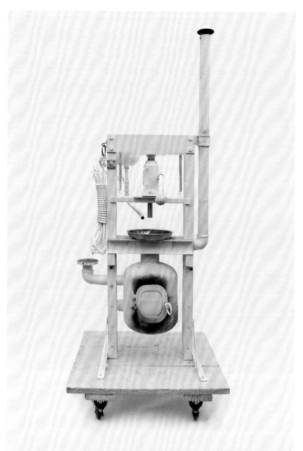

联合国估计全球海洋含有大约一亿吨塑料,而从太平洋环流中取样调查的结果表明,塑料与浮游生物之比是6∶1磅。随着我们社会的消耗增多,塑料的含量还在提高。《海洋椅》完全用从海洋中回收的塑料来制成。Swine 工作室与基耶伦·琼斯合作创造了一些装置来收集海洋碎屑,然后将它们处理成一系列的产品。

The UN estimates the world's oceans to contain some 100 million tons of plastic, samples from the Pacific gyre have shown ratios of 6 pounds of plastic to 1 pound of plankton. As our society's consumption grows, the concentration of plastic increases. The "Sea Chair" is made entirely from plastic recovered from our oceans together, Studio Swine and Kieren Jones have created devices to collect and process marine debris into a series of products.

海洋椅,2011
Swine 工作室 / 亚历山大·葛罗夫斯、阿祖撒·穆拉卡米、基耶伦琼斯(英国)
综合材料
© Studio Swine

The Sea Chair Project, 2011
Studio Swine /Alexander Groves , Azusa Murakami, Kieren Jones (UK)
Mixed Material
© Studio Swine

以天然竹丝为材料，利用传统手工穿丝制作而成；在设计上，作者做到物尽其用，顺从竹丝的天然柔软性，将竹丝的自然特征很好地融入产品造型设计中，同时也将竹子天然透光性的特质表现得淋漓尽致；在结构上，实现了零五金件的设计方式；整个产品将竹丝工艺、竹圈工艺及全竹工艺都结合在一起。

Natural bamboo stripes are taken as the materials, and the works is made with conventional strip threading handwork. The author trys to make full use of the materials in design, and follow the natural flexibility of the bamboo stripes so as to integrate the natural properties of the bamboo stripes into the model design of the works. Meanwhile, the special property of bamboo, that is natural transparency, is fully demonstrated. The design method for hardware is embodied in structure. The whole works integrates bamboo strip technology, and bamboo circle technology as well as whole bamboo technology.

又见炊烟，2012
杨剑（中国）
竹丝
© 十竹九造

Smoke Spiraling Again , 2012
Yang Jian (China)
Bamboo Stripes
© MADE BAMBOO

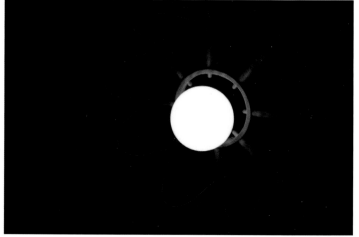

作者对人类无止境地对地球资源的攫取深感忧虑，对著名科学家霍金对未来的担忧表示认同，并以此为依据进行创作，且想通过此作品唤起人们对地球的忧患意识和共鸣。因为这是一个很严肃的话题，从每个国家，从我们每个个体的人做起，让我们一起携起手来共同努力去避免霍金先生的预言成为事实。我们的行为决定着地球的未来，地球的未来又影响着我们现在的生存状态，因此设计者认为到了人类应该真正思考并严肃对待这一问题的时刻了，希望此设计能达到这一效果。

The works is created based on the deep worries about the endless snatch of the resources of the earth by man, and recognition of the concerns about future of Hawking, a renowned scientist. The works is intended to awaken the awareness of unexpected development and resonance among people. As it is a quite serious topic, we shall start from each of us in each country, and jointly strive to prevent the prophesy of Mr. Hawking from becoming reality. Our act determines the future of the earth, which in turn influences our current living conditions. Thus the author thinks it is due time for man to carefully think about and serious treat such issue, and hope the design can achieve such effect.

无题，2012
马泉（中国）
平面设计
© 马泉

Untitled, 2012
Ma Quan (China)
Plane Design
© Ma Quan

波 , 2011
费边·温克勒、香农·麦克伦 (德国 / 美国)
印刷、刻录机、DVD 光碟、唱机、混声器
© 费边·温克勒、香农·麦克伦

Waves, 2011
Fabian Winkler, Dr. Shannon McMullen (German/USA)
Framed Prints, Records, DVDs, Record Player, Sound Mixer
© Fabian Winkler, Dr. Shannon McMullen

虽然在某些特定时刻，我们通过图片的分享使人们对自然环境留下深刻的印象，而这个项目用声音解释了自然环境中长期的气候效应。这是一个可以记录波浪信息的装置，具体是从艺术与科学两种不同的视角观测美国的密歇根湖。在卡里·特洛伊博士以及普渡大学一位研究流体动力学的专家的支持下，我们测量了两个不同类型的波浪，主要是测量湖内水体随波浪起浮而发生的温度变化。麦克伦和温克勒将观测所得到的数据转化成一个数学波形的形状，可以用于刻录黑胶唱片。Mat 实验室还可以制作成自定义的脚本，将数据转化成独立的声波。所记录下的水波形态被压成微型的格式，刻制成黑胶唱片上的凹槽，以此播放出声音。耳机中传来的共振不仅反映了水体运动的特征，还暗示我们该水体所在的密歇根湖的环境特点（如温度、季节、气候变化）。黑胶唱片成为回放和储存的媒介，将水波浪转化成声波的同时，储存所感应到的水体信息。黑胶片上的凹槽使得原始观测数据可以被倾听，而所传达出的声波系统反馈了科学研究、环境意识和艺术实验三个领域的交叉与互动。最终的黑胶片将包含所有大湖的水体信息。

While we often share images to give an impression of a natural environment at a particular moment in time, this project interprets a natural environment through longer term climatic effects realized as sound. The result is Waves Records, which brings together scientific and aesthetic understandings of Lake Michigan (USA). With the scientific support of Dr. Cary Troy, an expert in fluid dynamics at Purdue University, we were able to measure two different types of waves: internal waves measured below the lake's surface by temperature changes in the water. McMullen and Winkler translated this data into a mathematical waveform whose shape could be used to create grooves on a record. A custom script in MatLab turned data points into individual samples of an audio waveform. The waveforms of the recorded water waves were then pressed in microscopic format as grooves on a vinyl record that can be played back as sounds. The resonance heard through the headphones reflects water movement created by environmental conditions (e.g. temperature, season, climate change) specific to Lake Michigan. The vinyl record thus becomes both a playback medium for the sonified water waves and a storage medium for the sensor data. The grooves allow the original data to be heard and contemplated as soundwaves floating at the intersection of scientific research, environmental awareness and artistic experimentation. Eventually Waves Records will consist of a set of recordings from all of the Great Lakes.

机场用残疾人登机轮椅长久以来就存在着诸多问题，比如由于轮椅传统的金属材质会影响对残疾人乘客的安检，所以在通过安检门时必须由专人辅助更换轮椅，传统轮椅与登机轮椅尺寸存在较大差异性，不能直接登机、舒适度不够等一系列问题，使残疾人乘客的登机过程十分不方便。此项设计是"清华波音联合创新实验室"的设计项目，旨在通过好的设计改善残疾人乘客的登机体验。此设计通过全合成材料的应用，避免了在使用者通过安检门时需要更换轮椅的繁琐过程，加之轮椅的尺寸在满足舒适度同时满足登机的尺寸要求的情况下可直接登机，进一步简化登机过程，使得残疾人乘客的登机体验得到全面提升，达到舒适、快捷、易用等多方面要求。

There have been many problems with the boarding airport wheelchairs for the disabled for long. For example, the conventional metal materials of the wheelchair will influence security check of the disabled passengers. Thus a series problems such as special personnel must be arranged to aid them change the wheelchair at the security check door, and the great differences in size between conventional wheelchair and boarding wheelchair, failure to board the plane directly, and insufficient comfort etc. have caused great inconvenience to the disabled passengers during boarding. Such design is the design project of "Tsinghua Boeing United Innovation Lab", which is intended to improve the boarding experience of the disabled passengers through good design. The design has avoided the complex and tedious course of changing wheelchair through the security check door for the users through application of compound materials. In addition, the sizes of the wheelchairs can meet the requirements for the size for direct boarding while meeting the requirements for comfort, thus further simplifying the boarding course, and comprehensively improve the boarding experience of the disabled passengers to meet the comprehensive requirements such as comfort, rapid, convenient for use etc..

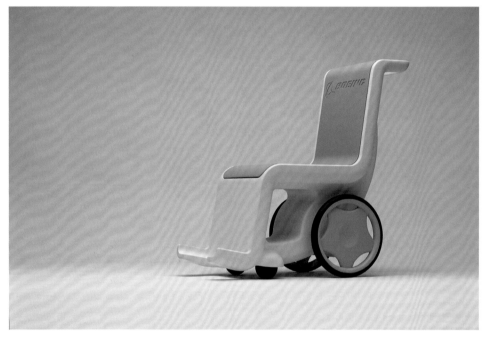

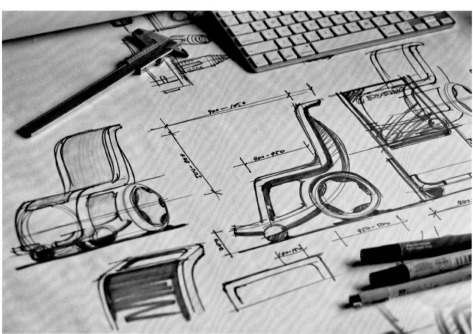

机场轮椅，2012
张雷、马赛、严扬、刘凯威（中国）
合成塑料
© 张雷、马赛、严扬、刘凯威

Airport Wheelchair, 2012
Zhang Lei, Ma Sai, Yan Yang, Liu Kaiwei (China)
Composite Plastics
© Zhang Lei, Ma Sai, Yan Yang, Liu Kaiwei

该雕塑是以胰岛素晶体结构（中国最接近诺奖科技成果）为原型，将胰岛素的分子结构与艺术再造形成雕塑的核心造型。分子结构模型的支架设计为优美的艺术造型，使雕塑更具观赏性和艺术性。中间的分子结构给人以破壳而出的形象效果，现代科学的不断创新已成为推动人类进步和社会发展的不竭动力。但科技也是双刃剑，雕塑在展示创新意境的同时，也体现了现代科技的快速发展使人类的欲望越来越膨胀，我们的地球家园正在被人类无止境的掠夺伤害着，带来无尽的创伤。外观球形的不完整感觉即表达了此理念。

The crystal structure of insulin (the Chinese technical achievement that approaches Nobel Prize the most) is taken as the prototype of the sculpture, and the molecular structure of insulin and artistic recreation form the core model of the sculpture. The support of the molecular structure model is designed with a nice artistic model to improve the appreciation value and artistic properties of the sculpture. The molecular structure inside brings about the image of bursting from shell, and the constant innovation of modern science has become the inexhaustible power to promote human progress and social development. However, science and technology are a double-edged sword, that is the sculpture has shown the increasingly expanding human desire due to rapid development of modern science and technology while displaying the innovative conceptions. Our earth is being plundered and damaged by man endlessly, and suffers from indefinite injury. The incomplete external sphere conveys such concept.

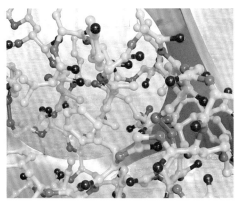

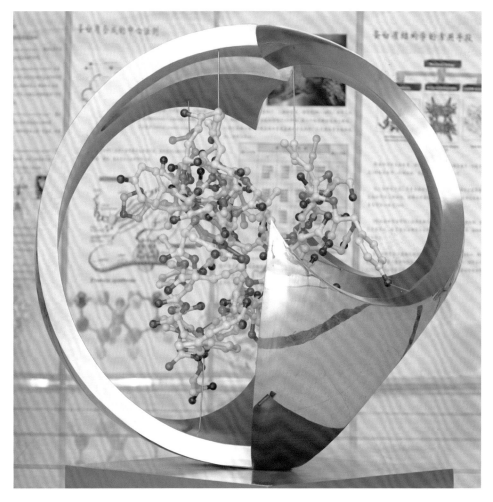

创，2011
汪会盛、王永刚（中国）
不锈钢
© 汪会盛、王永刚

Creation, 2011
Wang Huisheng, Wang Yonggang (China)
Stainless Steel
© Wang Huisheng, Wang Yongang

生物恋——器官工艺，2011
维罗妮卡·兰内尔（德国）
手吹玻璃、脱胶蚕丝
© 维罗妮卡·兰内尔

Biophilia—Organ Crafting, 2011
Veronica Ranner (German)
Hand Blown Glass, Degummed Silk
©Veronica Ranner

作品旨在挑战设计在响应尖端生物技术世界中新兴需求时的角色。作者提出了未来世界可能的新视像，运用设计这种介质来鼓励大众从文化、道德和伦理问题方面重新评价自己与科学技术的关系。生命与非生命、原生与人工、天然与制造之间模糊的界线方面已经产生了许多范式的转变。想想器官移植吧，仅仅几十年前，它还被视作是不合自然规律的事情，如今在我们的日常生活中早已司空见惯。然而，科学家与政治家们在将来的技术应用方面作出的决策对人类命运将会产生重大的影响。合成生物学方面范式的转移甚至还能对我们与生俱来的"生物恋"观点提出挑战，这个观点是埃里希·弗洛姆提出的"爱慕一切生命"的理论——比如说，如果可以用经过基因改造的蚕来代替机器为你编织一个辅助心脏运作的生物支架，那你愿意选哪一种？

合成生物学与组织工程学的进步有可能改变我们对自然环境的看法，还可能改变我们融入自然环境的方式。"生物恋"中的器官工艺部分反映了五千年悠久的育蚕历史传统。过去的一百年间，家养桑蚕的功效提高了200多倍，靠的就是选择育种。自2008年蚕的基因被解码后，人们已经能够用蚕来编织可以生物降解的支架，用于器官、组织移植和生物传感器安装，甚至还能编织从"硬件"到新颖的"湿件"这类产品，而不再用它们来结出丝茧。由于心血管疾病是全球头号杀手，心脏的捐赠将会日益稀缺。丝织心脏支架可以利用取自患者的细胞来"播种"，然后让它们生长成完全独立的器官，而不会造成任何排斥。但是，在处理生命材料时是否需要更加人性化的方式？这对我们与无生物界的关系又会产生什么样的影响？

The aim of my work is to challenge the role of design in response to emerging needs in a world of state of the art biotechnologies. The author proposes alternative views of potential futures using design as medium to encourage the public to reassess their relationship with science and technology with regard to cultural, moral and ethical issues. Many Paradigm shifts in the blurry line between the living and non-living, the organic and the artificial, the natural and the engineered have already taken place. Thinking about organ transplantation, which was seen as unnatural just a few decades ago, is common practice in our everyday lives. But the decisions made by scientists and politicians about the future application of technologies will shape in no small part human destiny. New paradigm shifts regarding synthetic biology could even challenge our inherited view of "Biophilia"—Erich Fromm's thesis on the affection towards everything living. If genetically modified silkworms would weave the scaffold for your donor heart for example instead of a machine—what would you prefer?

Advances in synthetic biology and tissue engineering might provide alternative views on how we see and embed ourselves in the natural environment. Biophilia—Organ Crafting refers to the 5000 year old tradition of silkworm breeding. The domesticated silkworm Bombyx Mori's efficacy has been optimized by more than 200 times in the last 100 years, just by selective breeding. Since the silkworms' genetic information was decoded in 2008, it could be altered so the silkworms weave biodegradable scaffolds for organs, tissues, biosensors and even products instead of their cocoons from "hardware" to novel "wetware". As cardiovascular diseases are globally number one cause of death, we will face an increasing scarcity of donor hearts. The silken heart scaffolds could be seeded with cells, gathered from the patient, to then grow a wholly individual organ without rejection as consequence. But does dealing with living material require a more humane way of production? And how would this impact on our relationship with the inanimate world around us?

在《生物恋——存活组织》这件作品的中,作者更进一步地思索了体外人工组织潜在的用途。我们应该如何使用这种合成的人体肌肉,它又能提供什么样的创新功能?当前我们面临着早产婴儿日益增多的问题——全球大约10%的婴儿都属早产,这种趋势还在继续扩大,而早产可能对婴儿产生长期的不良后果。引起这种情况的因素有可能是越来越多的不育症、慢性应激以及孕产期的推迟。近年来新生儿护理方面的进步能够保全300克左右——才一片黄油那么重的婴儿的生命。合成皮肤的设计还能提高早产婴儿的存活概率,因为它能提高皮肤的接触率。而在这类技术进步中,我们又会面临什么样的伦理问题和情感问题呢?你如何衡量它们在医学方面可能带来的益处?你又如何划分生命的界线?

With Biophilia—Survival Tissue, the author speculated further on the potential application of keeping such created artificial tissue in vitro. How may we use this synthetic human flesh and what innovative functions would this provide? We currently face an ever-increasing numbers of premature births– around 10 percent of births worldwide are premature. The tendency is growing and the long-term consequences for the infants alarming. Suspected factors are increasing fertility treatments, chronic stress and delayed motherhood. Recent advances in neonatal care preserves the lives of babies around 300 grams, which equates to a piece of butter. A design response with synthetic skin could even improve an extreme premature infants' survival chance due to the increased skin contact, but what ethical and emotional problems will we face through the progress of such technology? How do you measure potential medical benefits? And Where do you draw the boundary of life?

生物恋——存活组织,2011
维罗妮卡·兰内尔(德国)
手吹玻璃、快速成型物件、硅、树脂玻璃、脱胶蚕丝
蚕丝印刷、在真丝织物上的数字印刷
© 维罗妮卡·兰内尔

Biophilia—Survival Tissue, 2011
Veronica Ranner (German)
Hand Blown Glass, Rp Objet, Silicone, Perspex, Degummed Silk
Silk print: Digital Print on Habotai Silk
©Veronica Ranner

高通量微阵列生物芯片扫描仪获得中国政府设计创新最高奖"红星奖"。该仪器是为集成医疗（包括预测、预防和个体化医疗）领域开发的创新性产品和服务，应用生物芯片（包括基因、蛋白、细胞芯片等）技术，对未来个体化医疗提供技术服务。高通量微阵列生物芯片扫描仪提高了产品自动化和扫描通量，可搭载一个一次能装载24片生物芯片的自动装载器和一个读码器。精心设计的芯片卡盒能为使用中的芯片增加更多额外的保护。该产品拥有很高的灵敏度和准确性，较宽的动态范围和线性度。专门设计的应用软件是集成扫描、图像处理、数据统计和分析、报表生成功能于一体的系统平台，功能更紧凑，提供用户自定义流程设置和分析界面。用户可以设计自己的工具来进行数据信号的处理和判别。主要应用领域集中在高、中、低密度生物芯片的双色或单色荧光检测分析上。包括临床检验，如自身免疫性疾病检测和细菌鉴定与耐药检测等；食品安全检测，如兽药残留检测和食品微生物检测等；生命科学研究，如DNA甲基化检测和基因表达谱分析等。

"High Throughput Microarray Labnochip Scanner" has won "Red Star Prize", the highest prize for design innovation of Chinese government. Such instrument is an innovative product and service developed for the field of integrated medical care (including prediction, prevention and individualized medical care), and provides technical service for future individualized medical care through application of labnochip (including gene, protein, and cell chip etc.). High throughput microarray labnochip scanner improves the automation degree and scanning throughput of the product, and can carry an automatic carrier that can loads 24 labnochips at one time and a decoder. The meticulously designed chip box can increase more extra protection for the chip in service. The product boasts quite high sensitivity and accuracy, wide dynamic scope and linearity. The specially designed application software is a system platform that integrated scanning, image processing, digital statistics and analysis, and statement generation with more compact functions, and can provide self-defining process configuration and analytic interface. The users can design their own tools to carry out disposal and judgment of digital signal with their own tools. The major application fields are double-color and single-color fluorescent testing and analysis of labnochip of high, medium and low density, including clinic testing, for example, testing of own immunity disease, and germ identification and drug resistance testing etc.; foodstuff safety testing, for example, testing of residues of animal medicines, and food microorganisms etc.; life science research, for example, testing of DNA methylation testing and gene expression profiles etc..

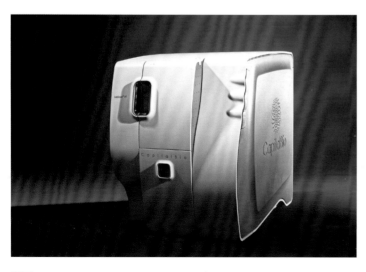

高通量微阵列生物芯片扫描仪，2010
赵超（中国）
综合材料
© 赵超

High Throughput Microarray Labnochip Scanner, 2010
Zhao Chao (China)
Composite Materials
© Zhao Chao

作品通过从李天元的血液里提取的 T 细胞，制作了三种状态：一是自然状态下的 T 细胞；二是向细胞里面添加了两种抗体 CD3、CD28，在这两种抗体的刺激下，T 细胞被激活，可看到的细胞会变大，表面突起增多，对外来刺激的反应增强；三是细胞逐渐死亡的状态，用 Fas 这一试剂处理细胞，这种试剂能够引起细胞凋亡（细胞凋亡是指为维持内环境稳定，由基因控制的细胞自主的有序的死亡）。作品通过对生命最基本的元素细胞生长被干预，变异抗争到有序的死亡，揭示了超出了人的意识之外的生命生存的智慧。

The works makes out three forms through abstracting T cells of Li Tianyuan. Firstly, T cell in natural state; Secondly, two antibodies such as CD3 and CD28 are added into the cells. T cell is activated with the irritation by the two antibodies. The cell will grow larger with more surface swellings, and the reaction toward foreign irritation will grow stronger. Thirdly, gradual death of cells. Cells are treated with the reagent Fas, which can lead to cell apoptosis (cell apoptosis refers to the orderly voluntary death of cells controlled by gene to maintain stability of internal and external environment). The works reveals the wisdom of life and survival beyond human awareness through interention, variation, fighting to orderly death of the growth of cells as the most basic element of life.

我的细胞，2012
李天元、俞立（中国）
彩色图片
© 李天元、俞立

My Cell, 2012
Li Tianyuan, Yu Li (China)
Color Picture
© Li Tianyuan, Yu Li

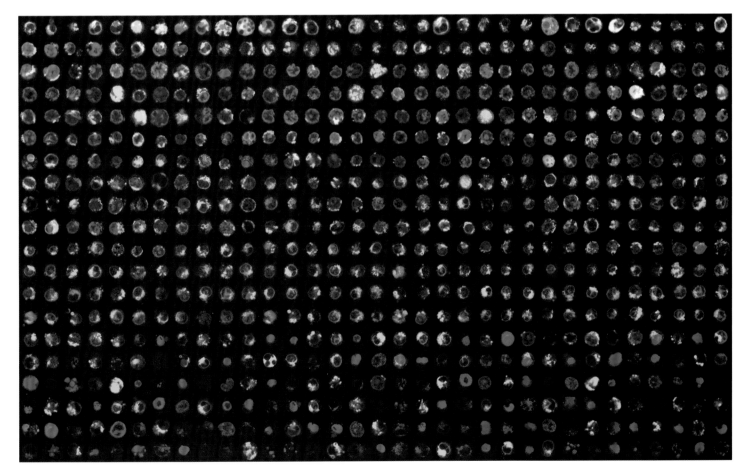

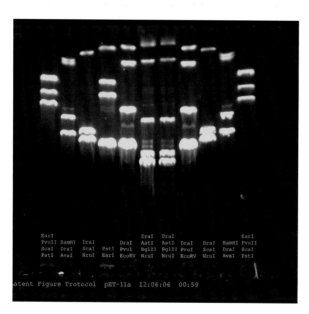

潜藏图形协议，2007
保罗·梵努斯（美国）
生物技术装置、六个灯箱
© Paul Vanouse

Latent Figure Protocol, 2007
Paul Vanouse (USA)
Live Performative Biotechnological Installation, 6 Lightboxes
© Paul Vanouse

这件作品运用了 DNA 分析中常用的凝胶电泳工艺，但却不是用它来抽象地描绘未知的 DNA 样本。作品颠覆了这一工艺作用，故意用它制作出某件已知 DNA 样本的结合模式，从而创造出一种意味深长的主题，也创造出了一种比喻意象。如此可见，梵努斯是想让 DNA 的科学权威降级为区区一幅图画。图元似的结合模式如今象征着一些可视的主题，比如 "ID"、"O1"、"©" 符号、母鸡与鸡蛋或是海盗的骷髅标志。于是这些在认证实验室标准工具的辅助下生成的图形便不再具有任何科学权威，而成了某种社会文化评论。在这道工艺过程中，主题只能缓慢、渐进地生成，就像拼图游戏一样：一旦它们的意义被"解码"，所有其他解读便都不见了。

The Latent Figure Protocol makes use of the process of gel electrophoresis that is employed in DNA analysis. Instead of the customary abstract portraits from an unknown DNA sample the procedure is subverted and purposefully produces banding patterns from a known DNA sample, thus creating significant motifs. Here, the process is used to create figurative images from a known DNA instead of the customary abstract portraits from an unknown DNA sample. Vanouse thus wants to downgrade the scientific authority of a DNA fingerprint to the status of a portrait. Pixel-like banding patterns now symbolize visual motifs such as ID, O1, the © symbol, hen and egg or the "skull & crossbones" pirate design. These diagrams, which are created with the aid of certified laboratory standards, thus have no longer any scientific authority. They rather become socio-cultural commentaries that co-opt the biochemical protocol. In the process, the motifs are created only slowly and progressively like picture puzzles: Once their significance is "decoded," all other potential interpretations are blinded out.

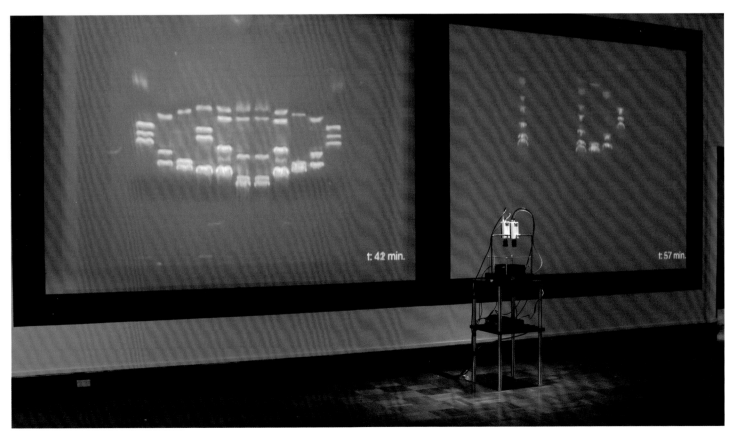

现在是2010年了,那么女人为什么还为月经而苦恼?

作为一名女性艺术家,作者想要解决这个让她很感兴趣的问题。

当20世纪60年代市面上开始出售避孕药时,女人每个月还必须在经期的一周中停药。这主要是因为医生们认为服药者对不来月经会感到忧心忡忡、难以接受。自那以后又过了50年,现代技术有了更多成就——太空旅行、手机、互联网、克隆以及基因改造食品。可女人还在出血。最近已经开发了"安雅避孕药"和"四季避孕药",这类将月经减少到每年零到四次的新药却没有得到广泛采用。

月经对人类究竟有什么生物、文化和历史方面的意义?谁会选择来月经?来月经的方式又是如何?

"月经制造机"是一套模拟下腹部血液分配的机械结构和电极装置。通过佩戴这个装置,男性也可以感受到经期的痛苦与流血的状况。这个项目在设计和研制上得到伦敦帝国理工学院医学系教授扬·布罗森的支持。《月经制造机——隆之选择》这套音乐视频讲述的是一名有易装癖的日本男孩隆在某一天选择来月经,进而从生理上装扮成女性。他制造了这台月经体验装置,把它安在身上,以此来满足他理解女性朋友们在月经期间真正感受的愿望。

It's 2010, so why are humans still menstruating?
As a female artist, the author had one intriguing question she wanted to solve. When the contraceptive pill first became commercially available in the 1960s, it was deliberately designed to have a pill-free, menstruating week every month. This was mainly because the doctors felt that users would find having no periods too worrying and unacceptable. 50 years have passed since then, and modern technology has accomplished even more—space travel, mobile phones, internet, cloning and genetically modified foods—but women are still bleeding. New pills such as Anya and Seasonique that reduce the frequency of menstruation to 0 or 4 times a year have recently been developed, but they still don't seem to be widely used.
So what does menstruation mean, biologically, culturally and historically, to humans? Who might choose to have it, and how might they have it?
The "Menstruation Machine"—fitted with a blood dispensing mechanism and electrodes stimulating the lower abdomen—is a device which simulates the pain and bleeding of a 5 day menstruation process. The actual machine was designed and developed with research support from Professor Jan Brosens at the Department of Medicine, Imperial College London.
The music video of "Menstruation Machine, Takashi's Take" features a Japanese transvestite boy Takashi, who one day chooses to wear Menstruation in an attempt to biologically dress up as a female. He builds and wears the machine to fulfill his desire to understand what the period really feels like for his female friends.

月经制造机——隆之选择,2010
Sputniko!(英籍日本人)
装置
© Sputniko!
作者及东京柏汤屋艺术馆友情提供

Menstruation Machine, Takashi's Take, 2010
Sputniko! (Japanese British)
Device
© Sputniko!
Courtesy the artist and SCAI THE BATHHOUSE, Tokyo

作品灵感来自关于灰雀灭绝的一则报道。更多动物的灭绝与人类的劣行有关已经成为一个不争的事实。科技带给人类更大的自信，似乎我们已成为上帝一般，可以创造生命，同时也可以轻而易举地毁灭生命。然而每个生命本身自有它自身的存在方式与价值。科技的进步寻求的是一种人与自然更为和谐的生存方式，我们从自然中来，到自然中去，人本身跟自然界的一切是一个共同体，我们息息相关。所有的生命之间皆是一种轮回的存在，唯有以一种博爱与谦卑的心态来使用我们的科技面，去面对生命，才能实现一种真正的和谐。

The inspirations of the works come from a report on the extinction of sparrow. The relation of the extinction of more animals and the misdeeds of man has become an undisputable fact. Science and technology have brought about greater self-confidence to man and we seem to have become God that can create life, or easily destroy life at the same time. However, each life has its own means of survival and value. Scientific and technological progress is intended to seek a more harmonious means of survival between man and nature. We came from nature and will go into it as well, for man and all the things in nature are a closely related whole body. There is rotational existence among all the lives, and a real harmony cannot be realized unless our science and technology are used with universe fraternity and a humble attitude.

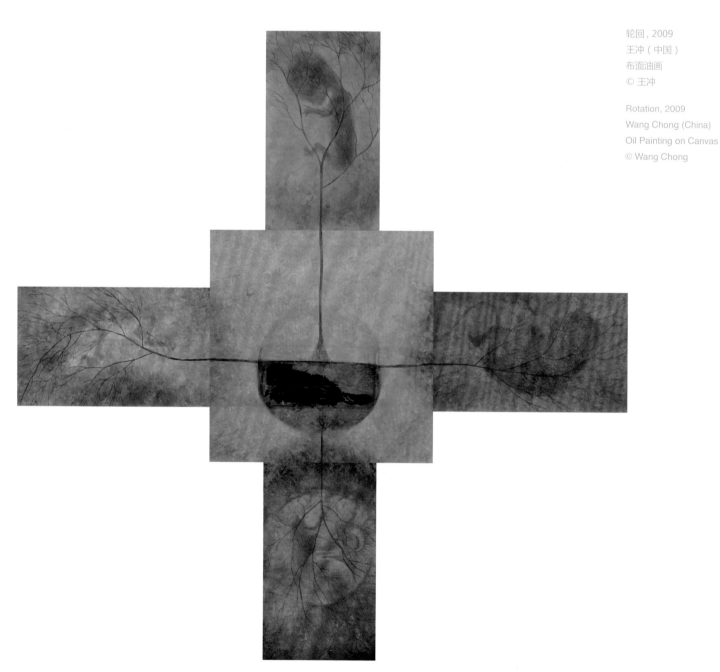

轮回, 2009
王冲（中国）
布面油画
© 王冲

Rotation, 2009
Wang Chong (China)
Oil Painting on Canvas
© Wang Chong

作者能够娴熟地运用天然大漆、蛋壳镶嵌等漆艺技法表现自然，表现瑞雪带来的祥和宁静的气氛。

在画面中，天然大漆调和色粉在色彩的表现上独具魅力。运用铝粉、蛋壳、色粉表现雪天。蛋壳镶嵌工艺将飘雪、积雪、冰雪及树枝干上的雪等表现得出神入化。

The author can express nature through proficient use of lacquer skills such as natural lacquer and egg shell marquetry etc. to express the peaceful and quite atmosphere of auspicious snow.

The ready-mixed color powder of natural lacquer has unique charm in expression of colors in the picture, in which aluminum powder, egg shell and egg powder are used to show a snowy day. The egg shell mosaic has presented perfect demonstration of the flying snow, accumulated snow, icy snow and the snow on the branches etc..

瑞雪祥琪, 2009
白小华（中国）
大漆、蛋壳
© 白小华

Auspicious Snow, 2009
Bai Xiaohua (China)
Chinese Lacquer, Egg Shell
© Bai Xiaohua

音乐有一种整体的进化感，因为我们的感官会为其触动。细胞的有丝分裂诱发了这幅作品的中心结构。不过在这种形态中，我们看到的不是DNA，而是从斯科特·乔普林的《枫叶散拍》中节选的断章，有如一串缠绕的音符。这首乐曲在爵士乐的发展中影响深远。爵士乐的兴起是为了抚慰生命中的悲哀，是为了表达痛苦和战胜痛苦。希望它能像万灵的魔法石，化解古往今来代代相传的人间苦难。

There is a sense of evolution in the development of music that is holistic in that we are touched in all of our senses by music. The principle of mitosis and the division of a cell motivated the central configuration of this works. Within this form instead of DNA we see an extract from *Maple Leaf Rag* by Scott Joplin as an intertwining sequence of notes. This piece of music is one of the most seminal works responsible for quickening the development of Jazz. Jazz evolved to deal with life's tragedies, to express pain and to over come them. The author refers to Jazz here is a kind of panacea, a philosophers stone to counteract that Pandora's box of ancient inherited maladies known to us as the genes of the Morbid Map.

蓝调的诞生，2006
休·奥唐纳（美国）
数码设计
© 休·奥唐纳，波士顿大学

Birth of the Blues, 2006
Hugh O'Donnell (USA)
Digital Design
© Hugh O'Donnell, Boston University

这是为波士顿大学迈阿密生命科学大楼创作的 10 幅视频截屏墙画,采用了影片《聆听迈阿密》中的许多硅藻,画面右上方是微型海链藻线粒体基因组的形状,箭头表示基因的转录方向。蓝色箭头摹写了蛋白加密基因,黑色箭头摹写的是转移核糖核酸。背景中还有一些字母,它们的形状构成了预期的蛋白质,并依据质体转运序列的识别来分布其位置。

Features many of the Diatoms from the film *Hear of Miami* a 10 screen video wall production for the University of Miami Life Science Building. The image features a Gene Map (top right) of the Diatom T. Pseudonana diatom mitochondrial genome. Arrows indicate transcriptional direction of genes. Blue arrows depict protein-encoding genes. Black arrows depict tRNAs. B Also in the background there is the text beginning to form the proteins predicted to localize to plastid based on identification of plastid transit sequence.

生命之海,2011
休·奥唐纳(美国)
数码设计
© 休·奥唐纳,波士顿大学

Sea of Life, 2011
Hugh O'Donnell (USA)
Digital Design
© Hugh O'Donnell, Boston University

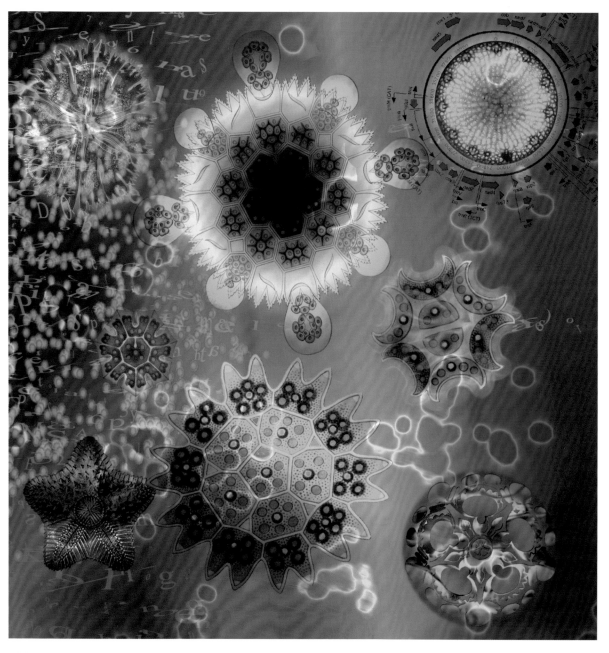

《生命轮回变奏》属于一幅更大的印制作品的一部分，它用手部的主动脉来描绘月球，而不采用环形山。这种月球概念被置于大麦哲伦星云中，也是中微子和新星辰的著名诞生地。

这套数码艺术组画是为纽约大学附属医院新建的心血管中心绘制的。作品的理念是激发心血管病患者们——新生的愿望。即，痊愈是理解自然的结果，也是与自然的过程缔结一种移情关系的进阶。之所以选择计算机辅助数码介质来作为这件艺术品的工具，是要让人类飞上月球的技术灵性也为医学与艺术的创新服务。

"Circle of Life Variation" is a design that is part of a larger installation of prints. It depicts the moon with the arteries of the hand instead of craters. This concept of the moon is paced in the Large Magellenic cloud a notable birthplace of neutrinos and new stars.

This series of digital site-specific artworks were created for the new Cardiovascular Center (CVC) at New York University Hospital. The concept for this work was based on inspiring the idea of new growth in the patients at the cardiovascular center. The artwork is intended to affirm that healing is a function of understanding nature and developing an empathic relationship with natural processes. The choice of computer assisted digital media as the instrument of art is in part an acknowledgement that the technical ingenuity that took us to the moon has also provided us with innovations for medical science and for art.

生命轮回变奏，2012
休·奥唐纳（美国）
数码设计
© 休·奥唐纳，波士顿大学

Circle of Life Variation, 2012
Hugh O'Donnell (USA)
Digital design
© Hugh O'Donnell, Boston University

作品是一个关于健康的可视化跟踪、测量与显示系统，研究动植物的健康行为并预测寿命。体现了健康的一个新度量：生物年龄指标。这决定了一个物体的生命长度，例如苹果、橘子或番茄。作品由一个传感器、自动监测及无线传输设备等组成。透过对有机生命体健康行为的长期监测，对及其微小的体力活动等进行运算，最终得出生物年龄和预期寿命的数据，显示在个人随身数码设备中。

"Memento Mori" is a health tracker and avatar display system that measures and visualizes the effects of health behavior to forecast how soon a person will die. Health is measured with a new metric: telomere length as an indicator of biological age. In calculus with life expectancy, biological age determines how much of a person's life remains.

The tracker is a sensor that automatically monitors and wirelessly uploads data about a person's health behavior, such as physical activity. It is as small, inconspicuous, and as easily carried as a coin. Upon tapping, biological and chronological ages and life expectancy are illuminated and visualized as concentric circles and individual numerical figures.

The display synthetically ages living organisms as avatars. Display avatars are embodiments of a person's chronological and biological ages. The display of a person who is "biologically young" shows little contrast between avatars as they develop, wither, and die in unison. The avatars of a person who is "biologically old" shows great contrast: the biological age avatar dies prematurely, just as the person will in life.

生物年龄追踪器，2012
麦克·亚普（美国）
装置艺术品
© 纽约视觉艺术学院（SVA）

Memento Mori, 2012
Michael Yap (USA)
Installation
© School of Visual Arts (SVA)

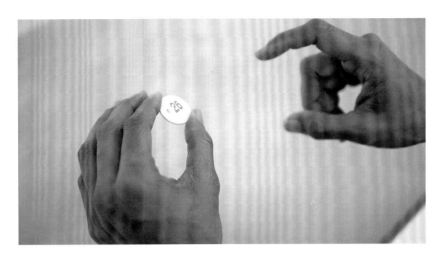

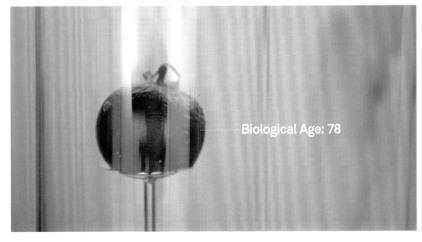

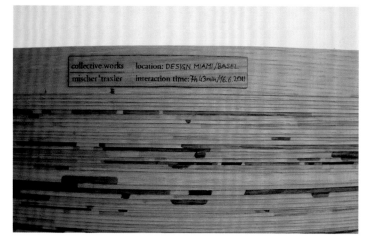

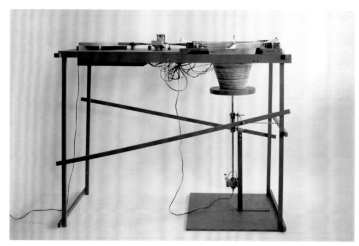

《集体作品》是这样一种生产过程：人们注意到它的生产机器时，机器才会完全发挥功能。机器对观众作出反应，把客流量转换成物体。随着它生产过程中人们对它投入的兴趣多少不一，它的产品色彩和尺寸也会变动不定。

一旦某个人走近前来观看机器，生产过程便开始了。一条24毫米宽的薄木板条经过一只胶水槽被抽出来，然后在20毫米厚的木制底座上慢慢盘旋而起，随着回转台带着木制底座向下移动，薄木板条便慢慢地编成了一只篮子。一旦有第二个人加进来观看这一过程，一支记号笔就会在薄木板上涂上一抹浅色。前来观看的人越多，启动的记号笔也就越多，每一支笔的色调渐次加深，最多的时候同时启动四支记号笔，把薄木板条涂得发黑。

这种互动是通过安装在机器框架中的传感器来实现的，它们能探测到观众的来往。篮子的高度由整个互动过程的时间长度来决定。驻足观看的人流越是多，篮子就会编得越高。

由于机器直接对每一名参观者作出反应，它的产品也就直接取决于每位观众。每一名参观者都在这件物品上留下痕迹，这样每只篮子都独特地记录下了人们对它的生产过程投入的兴趣。一只篮子，一件用来装东西的容器，成了它自己的数据集。如果没人对这一生产过程感兴趣，机器就会完全停止生产，也就编不出篮子来。这种生产方式可以被称作"应兴趣而生产"。

《集体作品》还质疑了人与机器之间的关系。观众被转换成了劳动者，尽管他们所付出的只不过是停下来观看的时间——然而我们最缺少的便是时间。通常工厂中的许多机器都只需要一些技术人员来监控它们的生产就行了，突然之间，一台机器竟然需要观众，才能生产出鲜艳明快的物品。

"Collective works" is a production process which is just fully functioning when people pay attention to the producing machine. Reacting to its audience, the process translates the flow of people into an object. The resulting outcome varies in colour and size just like the level of interest is varying during the time of production.

As soon as one person is coming close and looks at the machine, the production process is started: A wooden 24mm wide veneer-strip is pulled through a glue basin and slowly coiled up around a 20mm thick wooden base. Since the turning platform with the base moves downwards the veneer strip slowly builds up a basket. Once a second person joins to look at the process, a light tone colour is added via a marker onto the veneer. The more people come to look at the machine, the more markers are activated, each with a gradient darker tone. This goes up to four markers, at the same time, staining the veneer-strip black.

The interaction is possible due to sensors in the frame of the machine which detect the audience. Depending on the overall interaction time the baskets' height is defined. The more often somebody stops by to watch the process the higher the outcome gets.

The machine directly reacts to each observer and thus the outcome is as well directly depending on the audience. Every spectator leaves a mark on the object and therefore each basket becomes an unique record of the people's interest in the object's production. A baske—a vessel used to collect something becomes a collection of data by itself. If nobody is interested in the project, it stops producing at all and the final object just does not get made. This can be seen as "production on interest".

"Collective Works" also questions the relation between man and machine. The audience is turned into workers even tough their effort is basically just their time they spend with the machine—but time is what most of us lack somehow. Normally many machines in factories just need some technician to monitor the production and suddenly one machine needs some audience to produce colourful, vivid outcomes.

集体作品 2011
米斯切尔-特拉克斯勒工作室：卡特琳娜·米斯切尔、托马斯·特拉克斯勒（奥地利）
橡木薄板、记号笔
© Mischer'Traxler

Collective Works, 2011
Mischer'traxler Studio-Katharina Mischer, Thomas Traxler (Austria)
Oak Veneer , Marker
© Mischer'Traxler

源于树的灵感，2008 年至今
米斯切尔 - 特拉克斯勒工作室：卡特琳娜·米斯切尔、托马斯·特拉克斯勒（奥地利）
棉、色彩、胶水、树脂、橡木
© 米斯切尔 - 特拉克斯勒工作室

The Idea of a Tree , 2008-ongoing
Mischer'traxler Studio-Katharina Mischer, Thomas Traxler (Austria)
Cotton, Colour, Glue, Resin, Oak
© Mischer'traxler Studio

《源于树的灵感》是一道天然材料与机械流程相结合的自主生产过程。这一过程采用太阳能来驱动,每天通过一套机械装置将太阳光的强度转换成一件物品。

它的产品反映了当天不同的日光条件。就如一棵树一样,这件物品成了它自己创生的过程与时间的三维记录。作品中的"一号记录器"从日出时开始生产,到红日西沉时便停工。日落之后就可以"收获"完工的物品了。它慢慢地"生长"物品,通过一台上色装置和一只从胶水槽而抽出纱线,最终把它们绕在一只模子上。

产成品的长度和高度取决于当天日照的时间长度。卷绕每一层的厚度与颜色则取决于太阳能的多少(日光越多,绕层越厚、颜色越淡;日光越少,绕层越薄、颜色越深)。

这个过程不止对不同的天气条件作出反应,而且还对紧邻机器的周边环境中出现的阴影作出反应。每一朵云彩、每一道阴影对成品的外观都有重要的影响。该设计可以产生出多种物品。将天然要素引入系列生产的这种观念提出了审视区位的一种新视角。"工业化区位"所重视的并不是地方文化、手工艺或资源,而是生产过程周围的气候要素和环境要素。你能从这件设计产生的不同物品上分辨出它们的生产地点。例如在赤道上物品的高度/长度始终都一样,而在北欧和中欧,季节变化对物品的形状就会产生不同的影响。在雨水丰沛的国家里,物品的颜色更深、绕层更薄,而在阳光明媚的地区,物品颜色更淡、绕层更厚。对于我们而言,《源于树的灵感》是从理想主义的角度来看待机械如何与自然相结合才能生产出更了不起的成果——在自然的韵律之中沐浴日光、从事生产的工业场所。

"The Idea of a Tree" is an autonomous production process which combines natural input with a mechanical process. It is driven by solar energy and translates the intensity of the sun through a mechanical apparatus into one object a day.

The outcome reflects the various sunshine conditions that occur during this day. Like a tree the object becomes a three dimensional recording of its process and time of creation. The machine "Recorder One" starts producing when the sun rises and stops, when the sun settles down. After sunset, the finished object can be "harvested". It slowly grows the object, by pulling threads through a colouring device, a glue basin and finally winding them around a mould.

The length/height of the resulting object depends on the sun hours of the day. The thickness of the layer and the colour is depending on the amount of sun-energy (more sun = thicker layer and paler colour; less sun=thinner layer and darker colour).

The process is not just reacting on different weather situations, but also on shadows happening in the machine's direct surrounding. Each cloud and each shadow becomes important for the look of the final object. Various "The Idea of a Tree" —objects are possible. The concept of introducing natural input into a serial production process suggests a new way of looking at locality. This "industrialized locality", is not so much about local culture, craftsmanship or resources, instead it deals with climatic and environmental factors of the process' surrounding. On a series of objects you can tell somehow the place of production. On the equator, for example, the objects would always have the same height/length, whilst in North and Middle Europe, the seasons help shaping the objects. In countries with a lot of rain the objects would be darker and thinner whilst in sunnier regions the objects would be paler but thicker. For us "The Idea of a Tree" is an idealistic vision on how machines combined with nature can produce great results, an idea of industrial halls with daylight and manufacturing within natural rhythms.

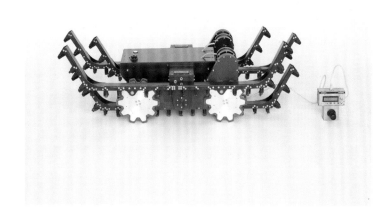

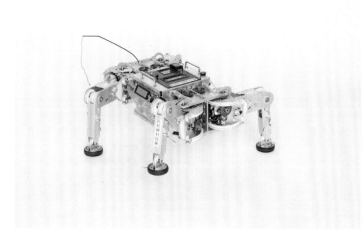

东京工业大学科学与艺术研究实验室的作品《创造流》为原创性的拼贴摄影作品，汇聚来自东京最先进科技实验情景的主题和图像。以挑战性的"Σ 部分"（即整体大于部分之和）为主题，该作品展示了多种美学价值以及技术物品作为艺术对象的巨大潜力。

作品系列的主题与本次作品展非常契合。自 2009 年推出后，《创造流》一直通过艺术创作，与科学家、艺术家和设计师合作组织创意座谈会、学术研讨会等，鼓励人们去思考科学、艺术及二者间的互动。

写在《创造流》网站上的话：

没有科学发现，我们不可能有目前的生活方式。

没有艺术的美，我们的生活将极端枯燥乏味。

科学和艺术都是我们所不可或缺的，但有时二者之间也会有冲突。就像水汇流到一起，会产生波浪和涟漪。我们将之称为"创造之流"。

在科学与艺术的碰撞与融合中，关注可持续发展的生态学正在兴起，这可以改变我们生活在其中的"城市生态"。

Collecting motifs and imagery from the cutting-edge experimental scenes in science and technology in Tokyo, "Creative Flow", Tokodai Science & Art Research LAB, presents original photographic collage works. Under the challenging theme "Σ parts" (= more than just parts), the works display various aesthetic values and the high potential of technological objects as pieces of art.

The theme which runs through this entire exhibition suits us very well. "Creative Flow" has been active since 2009 and has been encouraging people to think science, art and their mutual interaction for society by creating art, organizing Creative Cafés or discussion events, and seminars in collaboration with professional scientists, artists and designers.

From the "Creative Flow" website:

Without science's discoveries, we wouldn't be able to pursue our current lifestyles.

Without art's beauty, we wouldn't really be living.

Both are necessary, but they sometimes conflict, like water which flows together but with waves and ripples. This is what we call "Creative Flow".

Within the waves of science and art, energy and sustainable, deep ecology are flowing, and these can change the "city ecology" within which we live.

创造流，2012
科学与艺术研究实验室（日本）
综合实验材料
© 东京工业大学

Creative Flow, 2012
Tokodai Science & Art Research LAB (Japan)
Mixed Lab Material
© Tokyo Institute of Technology, Japan

微型智能机器人通过自身程序控制，在投影桌面自行运动，通过光电感应在桌面上留下机器人行走后的轨迹，犹如它在绘制一幅抽象的艺术画。我们也可以在桌面放置若干障碍物，让机器人绕着障碍物行走，并在行走过程中实现与观众的实时互动。

The micro intelligent robot moves on its own on the project desktop through its own controlling program, and the traces of the robot after walking are left on the desktop through photoelectric induction, with the robot seemingly drawing an abstract artistic picture. We can also place several obstacles on the desktop, and let the robot walk around the obstacles, and realize real-time interaction with the audience during walking.

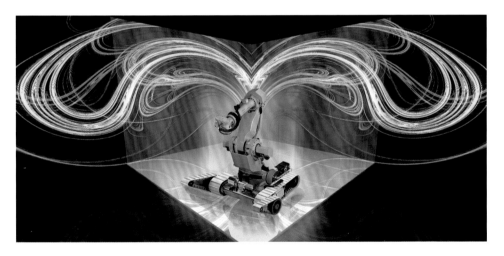

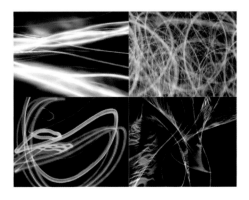

智能机器人行走绘图，2012
杨茂林、李喆（中国）
智能机器人、互动桌面、LED 灯光、亚克力装置
© 杨茂林、李喆

Picture of Walking Intelligent Robot, 2012
Yang Maolin, Li Zhe (China)
Intelligent Robot, Interactive Desktop, LED Light, Acrylic Device
© Yang Maolin, Li Zhe

这是一套双层玻璃墙面系统,用于大型建筑物,它有一层外部玻璃幕墙和一层内部玻璃表面,两层之间用气穴分隔开来,气穴中包含隐蔽的可变金属结构和通风装置,使这套系统能够自动调节热量的损耗和增益。

动态平衡墙面系统发挥了绝缘弹性体挠度高、功耗低的优点。丝带状设计能够开合自如,以控制透过墙面的太阳能量。

这种对环境条件作出自动反应的微调运动是通过一片人工金属单元实现的,这个单元由包覆在一只挠性聚合物芯体外面的绝缘弹性体构成。弹性体的膨胀和收缩使挠性芯体弯曲,而聚合物芯体端部的稳定轴确保系统随着弹性体的移动而产生平稳的运动。

绝缘弹性体两面均配有银电极,银能反射和散射光线,同时在弹性体上分配电荷,使之改变形状,从而对墙面及环境系统带来益处。

这种墙面受到生物系统中动态平衡的启发,通过对环境条件的自动响应来调节建筑物的室温。与普通墙面系统相比,它具有功耗低和精度极高的优点。由于表面材料同时也是系统的电机,因而它基本上能对任何一段墙面提供局部化控制,以进行温度调节,同时降低能耗和排放。

Contemporary double-skin glass facade systems for large buildings have both an exterior glass curtain wall and an interior glass surface that are separated by an air cavity. The air cavities often include shading mechanisms and vents that enable the system to automatically regulate heat loss and heat gain.

The Homeostatic Facade System advances double-skin technologies by taking advantage of the unique flexibility and low power consumption of dielectric elastomers. Its ribbon design opens and closes to control solar heat gain through the facade.

Automatically responding to environmental conditions, this highly tuned motion is achieved through a simple, elegant actuator. The actuator is an artificial muscle, consisting of a dielectric elastomer wrapped over a flexible polymer core. Expansion and contraction of the elastomer causes the flexible core to bend. A roller at the top of the polymer core ensures smooth motion as the elastomer moves.

The dielectric elastomer includes silver electrodes on both faces. The silver assists the system by reflecting and diffusing light, while distributing an electrical charge across the elastomer, causing it to deform.

Inspired by homeostasis in biological systems, the facade regulates a building's climate by automatically responding to environmental conditions. Its advantage over conventional systems lies in its low power consumption and superior precision. Because the surface material is also the motor, it essentially offers localized control along any segment of the facade, providing thermoregulation while reducing energy consumption and its associated emissions.

动态平衡墙面系统,2011
马提纳·德克尔、皮特·伊登
混合材料
© 纽约德克尔伊登公司

Homeostatic Façade System, 2011
Martina Decker, Peter Yeadon
Mixed Material
© Decker Yeadon LLC | New York

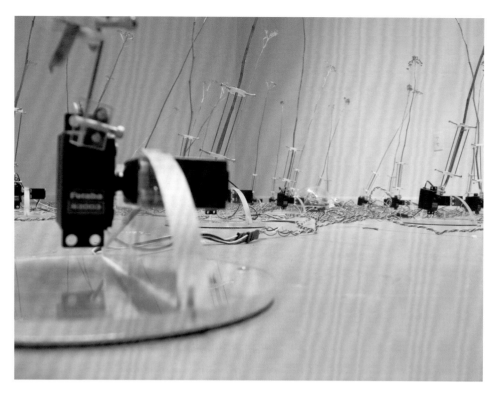

远程的风，2011
戴维·博文（美国）
铝、电子元件、干艾菊
© 戴维·博文

Tele-present Wind, 2011
David Bowen (USA)
Aluminum, Electronics, Dried Tansy
© David Bowen

这件作品由一系列 42 个维度的倾斜装置组成，它们连接着展厅中的干草茎，同时，室外的一只加速表上连接着同样干草茎。风吹过时，室外的草茎开始摇荡，加速表探测到这一动静，将它实时传输给展厅中的那组装置，于是展厅中的草茎也随着室外风一齐晃动起来。这件作品曾经安装在莫斯科艺术与科学空间实验室中，传感器则安装在明尼苏达大学可视化与数字成像实验室内。

This installation consists of a series of 42 x/y tilting devices connected to thin dried plant stalks installed in the gallery and a dried plant stalk connected to an accelerometer installed outdoors. When the wind blows, it causes the stalk outside to sway. The accelerometer detects this movement transmitting it in real-time to the grouping of devices in the gallery. Therefore the stalks in the gallery space move in real-time and in unison based on the movement of the wind outside. This piece was installed at Laboratoria Art & Science Space in Moscow and the sensor was installed at the Visualization and Digital Imaging Lab, University of Minnesota for an exhibition titledaesthetic data.

这件作品创造了第一场完全依靠自身能量来进行的自主式表演,借此重新思考了舞蹈表演旧有的生产模式。澳大利亚编舞师普鲁·朗进行了一种研究,想要把人类活动作为一种可再生能源。这件作品便开发了一套编舞系统,通过创造特定的技术,它使演员能够在表演的同时为灯光和音响供电,从而考察了人体与其环境之间的关系。朗与麻省理工学院媒体实验室的阿曼达·帕克斯合作发明了"智能"服装,它们能够自动收集舞蹈者在表演期间散发的能量。服装中内置的技术装置将舞蹈者运动时产生的动能转换为能量,存储在缝到服装中的小电池里,然后电池的能量又为音响系统供电,此外舞蹈者台上的单车还连接着低功耗高照度的LED照明网。演员(同时进行)摄入、花费、发电、响应、适应和生产。

This work re-thinks habitual modes of stage and dance production by creating the first autonomous performance that runs 100% on its own energy. Australian choreographer Prue Lang investigas human activity as a renewable source of energy. Developing a choreographic system in which the performers simultaneously power the light and the sound through the creation of specific technologies, the works examines the relationship between the body and its environment. Lang collaborated with Amanda Parkes from MIT Media Lab to invent "intelligent" costumes that could simultaneously harvest the dancer's energy during performance. Custom-built echnological devices convert kinetic energy generated by the dancers' movements into energy stored in small batteries sewn into their costumes. The energy from the batteries then power the sound system, while a low consumption high-luminosity LED lighting grid is hooked up to the dancer's onstage bicycle. The performers eat, expend, generate, respond, adapt and produce.

半透明的网，2011
普鲁朗（澳大利亚）、阿曼达帕克斯（美国）
视频
© 图片：Hillary Goidell

Un Réseau Translucide, 2011
Prue Lang (Australia),
Amanda Parkes (USA)
Video
© Photo: Hillary Goidell

凭借这件作品，作者想让参观者稍稍做一下白日梦，让他们把真人尺寸的铸造手想像成迈着编排的舞步而聚集起来的鸟群。作品本身是用八台单独的极简机器人、一台 mp3 播放器以及一台工业电脑而制成的，后者用来将硬件统一成一个群集的实体。

这些组件结合成了一种介质，后者随之又用来展示这件作品倏忽而逝的内容。机器短暂、有序的动作造成了一种对界线的模糊。雕塑与动画统一起来，机器人的手随着普罗科菲耶夫《骑士之舞》的音乐节奏而举起、放下。

With this work, the author wants to allow the observer to day dream a little. Allowing them to transform the life sized cast hands into a swarm that flocks together like birds in a choreographed dance. The installation itself is fashioned from 8 individual minimalistic robots a mp3 player and an industrial computer which is used to unite the hardware into a collective entity.

The combination of these elements becomes the medium that is in turn used to display the ephemeral content of the installation. The transient organized motion of the machines causes a blurring of boundaries. Sculpture and animation becomes united while the hands rise and fall to Prokofiev's "Dance of the Knights".

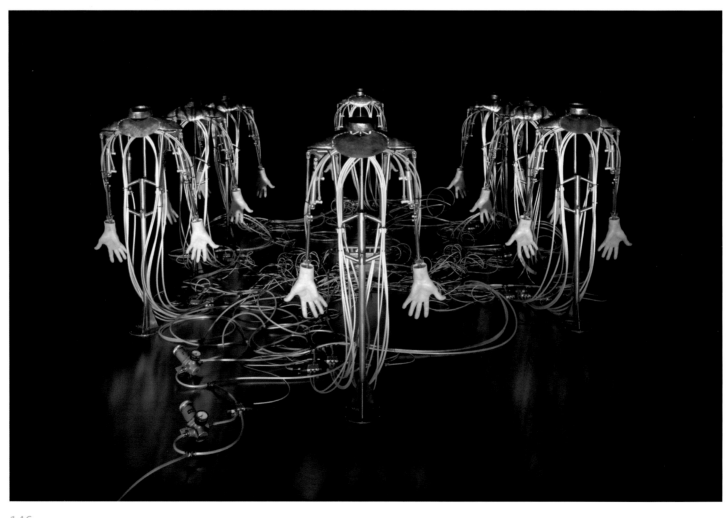

恶性循环，2012
彼得·威廉·霍尔登（英国）
复合塑料、计算机、压缩空气部件、Mp3 播放器
© 彼得·威廉·霍尔登

Vicious Circle, 2012
Peter William Holden (UK)
Composite plastic, Computer, Compressed Air Components, Mp3 Player
© Peter William Holden

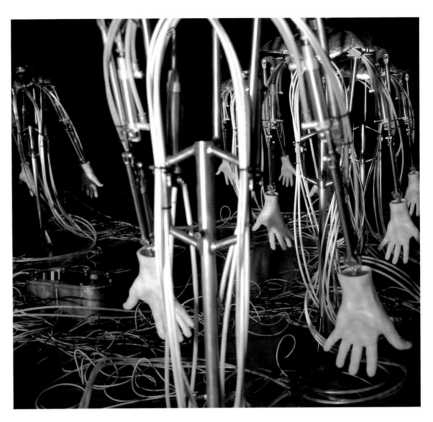
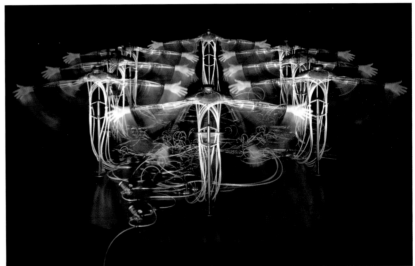
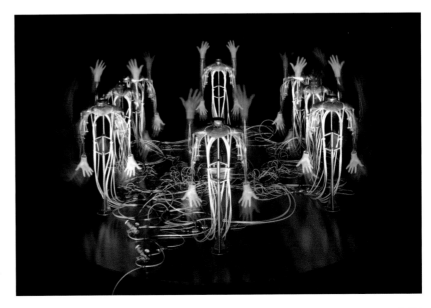

动态艺术要求作品在柔流或其他微力下进行缓慢、姿态优异、行径莫测的运动，在美感上增加了一个层次。作者选用了几何上可述的形状：线条、三角形、矩形、圆形及其组合变形。这种结合形状、运动和计算的追求组成了一种思维领域。作者概括出了若干条几何动艺的制约条件，"几何动艺"金属抽象造型。

Dynamic art requires the works to carry out gentle movements of superior postures and unfathomable course under the gentle flows or other minor forces, and add an aesthetic layer. The author has selected describable geometric shapes such as lines, triangular and rectangular, circles and composite deformation thereof. Such pursuit of shapes, movements and calculation forms a thinking field. The author has summarized several restrictive conditions for geometric dynamic art, and the metal abstract models of "geometric dynamic art".

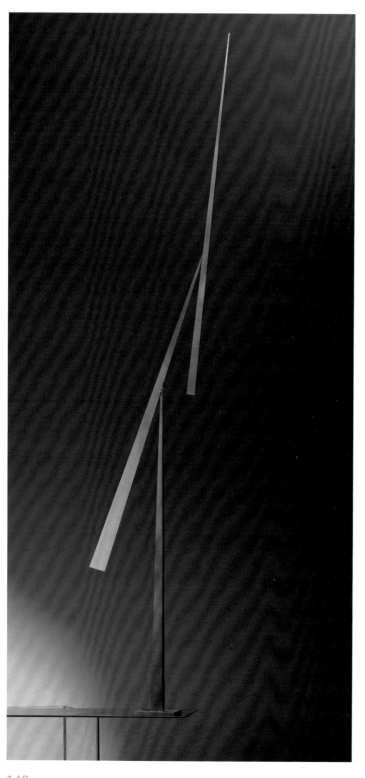

攀, 2007
郭慕逊（中国）
金属
© 郭慕逊

Climbing, 2007
Guo Muxun (China)
Metal
© Guo Muxun

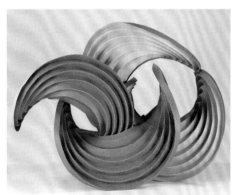

水彩画系列——折纸，2012
埃里克·德梅尼、马丁·德梅尼（美国）
Mi-Teintes 水彩纸
© 埃里克·德梅尼、马丁·德梅尼

Watercolor Series—Paper Folding, 2012
Erik Demaine, Martin Demaine (USA)
Mi-Teintes watercolor paper
© Erik Demaine, Martin Demaine

作品中每件雕塑都是多个组件的结合，每一片组件都是用圆形的纸手工折叠起来的，折痕都是同心的圆形。纸片依据它的折痕而形成自然平衡的造型。该作品探究这种弯曲的折叠方式能够产生什么样的形状，其结果可以应用于展开结构的制造以及自组装等问题的研究中。

Each sculpture is a modular combination of several interacting pieces. Each piece is folded by hand from a circle of paper with concentric circular creases. The paper folds itself into a natural equilibrium form depending on its creases. We are exploring what shapes are possible in this genre of curved-crease folding, which has applications to deployable structures, manufacturing, and self-assembly.

在类似的管状中求得每一个的形式不同,就像是人外形的相似又个性、修养、境界不同一样。探求这微妙的不同于作者有无穷的魅力,就像对待自己渴求懂得的未知一样。

Seek each different form in a tubular shape, just like people of similar appearances and different personality, cultivation and realm. Seeking such subtle differences is of infinite charm for the author, seemingly treating the desire to learn the unknown things.

管锥篇 隐语 II, 2011
白明(中国)
瓷土
© 白明

Chapter of Guanzhi, Insinuating Language II, 2011
Bai Ming (China)
Porcelain Clay
© Bai Ming

《绚丽太空》——用首饰(系列)造型语汇,表现太空中的神奇星云与太极的系列作品展现了宇宙中变幻莫测的星云,将千姿百态的星云转换成首饰的语言,是一种突破性的尝试。

在创作构想过程中,作者查阅了大量星云资料,并借助计算机辅助设计、绘制;同时,采用最新金属加工技术与器械加工成型。此项目是首饰造型新语言的一个尝试,也是艺术与科技紧密融合的成果。

"Brilliant Space"—It adopts jewelry (series) as the modeling language to express the fantastic nebula in the space and the Great Ultimate. The series works show the transient nebula in the universe, and convert the nebula of various forms into language of jewelry, which is an attempt to make breakthrough.

A large quantity of nebular data have been consulted during conception of the works, and reference has been made to computer-aided design and drawing. Meanwhile, the latest metal processing techniques and instrument processing and modeling have been adopted. The project is an attempt in new modeling language of jewelry, and an achievement made through combination of art and technology.

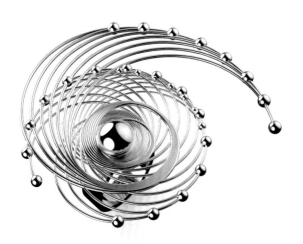

绚丽太空,2012
孙嘉英、陆永庆、刘旸(中国)
银、宝石等
© 孙嘉英、陆永庆、刘旸

Brilliant Space, 2012
Sun Jiaying, Lu Yongqing, Liu Yang (China)
Silver and precious stones etc.
© Sun Jiaying, Lu Yongqing, Liu Yang

《易经》有云："君子以同而异。"同中求异，异中见同，是古老文明的智慧结晶，是东方独特的思维艺术。作品中众多相同的个体如同现代社会中日渐趋同的芸芸众生，目光和思想指向不同的方向。而正是由于这样的同中之异，才使得众多个体组合成了中国传统玉璧的和谐形态，形成了包容众异的"大同"。

作品的制作工艺引入了现代数字喷蜡成形技术和立体建模技术等新工艺、新技术，使得古老的金属艺术与现代科学技术有机结合，体现出作者对现代金属艺术的理解。

As noted in *Classics of Changes*, "A gentleman varies for having common ground." Seeking differences in the common ground, and showing common ground in differences is the intelligent fruit of ancient civilization, and the unique oriental thinking art. The multiple identical individuals in the works are like the masses that tend to converge in the modern society, with the eyesight and thinking pointing to different directions. It is such difference in the common ground that enables multiple individuals to form a harmonious status like conventional Chinese jade Bi, and "general common ground" tolerating various differences.

New technology such as modern digital wax spraying and shaping technology and solid modeling technology etc. have been introduced into manufacturing of the works for organic combination of ancient metal art and modern science and technology, which shows the understanding of the author of modern metal art.

以同为异，2011
王晓昕（中国）
黄铜镀金
© 王晓昕

Common Ground Instead of Differences, 2011
Wang Xiaoxin (China)
Gilded Cooper
© Wang Xiaoxin

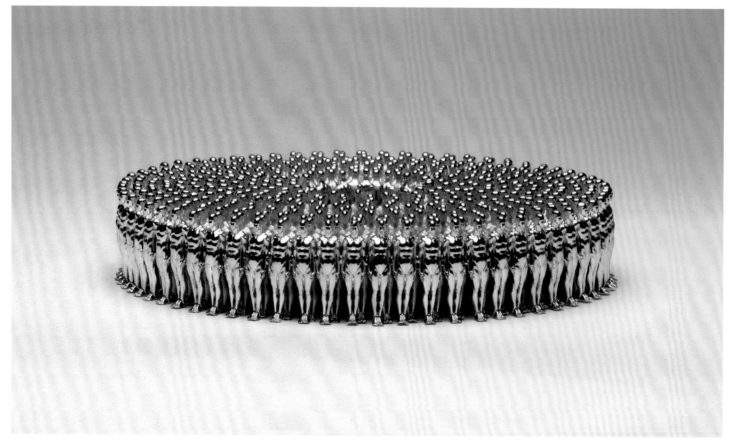

城市的不断繁殖体现了城市的生命和活力，全球化带来的美丽、炫目、光鲜、活力的外表，但在这外表下有着更多的生存焦虑，水资源危机、空气的污染、文化的全球化趋同直至特征的消失等。作品表现了人和城市这种若即若离、爱和恨交织的感受、城市给人虚无缥缈、晃动眩晕的感觉。

The constant propagation of cities has demonstrated urban life and vitality. Though globalization has brought about beautiful, brilliant, bright and vigorous appearance more anxiety about survival, water resources crisis and air pollution exist under the appearance. And cultural globalization converges until disappearance of features etc.. The works shows the lukewarm feelings with mixed love and hatred between people and cities, and the illusionary, shaking and dizzy feelings brought about by cities to people.

欲望城市——花系列, 2008
陈辉（中国）
不锈钢
© 陈辉

City of Desires—Flower Series, 2008
Chen Hui (China)
Stainless Steel
© Chen Hui

我们生活在宇宙中，强大的力量之间相互碰撞以形成无限空间的非静止性，这些形态相互吸引及排斥。作者试图解决缺少从正式仪式中脱离出来的不变性的情况，并在运动中体现自由的不对称性来述说未知的事物。对于我们人类而言，真理是没有途径到达的颂歌，而我们必须理解这一点，因此让我们在自然面前享受生命吧，尽管宇宙中存在着神秘的事物。在作者的艺术观点中，精神张力不仅重要而且必要。如果不去审视神灵、外部空间，我们无法理解，也无法看见、感知我们不了解的现存事物，则艺术仅仅是已知事物的领域而非产生思想的过程。

We live in a universe of forms, composing, drawing and rejecting one another because energetic forces collide for the non immobility of the infinite. The author trys to fix the absence of fixity, freed from formal ritual. The author craves to tell only the unknown, with a play of signs, form and volumes, free in the dissymmetry of their movement. For us, human beings, truth is a laud with no paths, and we have to be willing to understand that, so, in front of nature, let us enjoy being alive…in spite of the mystery of the universe that and tend to reveal…In the author's vision of art, tension towards spiritualy in not only essencial, but deeply necessary. Without our look towards the divine, the beyond, towards what we cannot comprehend but sense, what we cannot see but only perceive, towards that invisable that we don't know but exists nevertheless, art woved only be territory of the known and not of the generative process of the thought.

宇宙能量，2012
西蒙耐塔·格里亚诺（意大利）
帆布油画
© Simonetta Gagliano

Cosmic Energy, 2012
Simonetta Gagliano (Italy)
Oil of Canvas
© Simonetta Gagliano

作品为放置于一块铝板上并以图片印刷形式确定的照片雕塑。该艺术研究旨在从不同的观测点探究无限的可塑性潜力。

作品的标题为《空间分割》。作者实现了如同幻觉几何图形一样的三维制作，并具有明显的技巧及艺术与科学之间的创造性途径，而艺术成为其中的一个积极的因素及潜在的张力，科学展现了方法论上的严密及研究的计划性。

因此，通过采用与艺术创造的自由性，以及通过方法论和严密的科学程序获得的风格特征有关的模型，作品在不同层面寻求不断演变的动态过渡空间的过程中展现了艺术的诗意。

The works is a photo-sculpture placed on an aluminum plate and defined by a photographic print ; this artistic research aims to investigate the infinite plastic potentials, as suggested by the different points of observation. Under the title "Space Division", the author realizes a three-dimensional development, like geometries of illusions, with striking games and with a creative path fluctuating between art and science, art as a proactive element and underlying tension,science as methodological rigor and planned research.Therefore, through a matrix related both to the freedom of artistic creativity and to the stylistic features made through a methodical and rigorously scientific process, the author addresses the art on different levels of search for poetics defined as a dynamic transitional space of a process that is not stable but constantly evolving.

分割, 2011
阿费欧·蒙格里（意大利）
铝
© Alfio Mongelli

Divisione, 2011
Alfio Mongelli (Italy)
Alluminio
© Alfio Mongelli

互联网内在于全球化的逻辑之中。在当下全球化的浪潮里,互联网扮演着一个重要的角色,是全球化信息传播的核心手段。过去的生活方式、思维结构正在逐步变化,社会正在被消融、分解,新的社会模式正在形成之中。《正负互联网》应对这样的结构改变,以一种艺术的方式介入,并陈述对其问题的看法。一种以"网络意识作为新的意识形态"得到人们逐步认同,这种新的意识形态以网络主义(cybersm)为特征,它具有某种跨文化形态的色彩,并试图将民族、国家意识形态、文化等特殊属性整合一体。

以形态而言,作品以放射状居多,我们看到形态与形态之间是通过放射状两端的结来连接的,这个结将形体归类而又向空间敞开,而结与结之间从无限的形态向空间伸延,但同时每个结既是个体又是相互依存的。喻指某个特定的网络系统,由此形成一个庞大的无所不包的系统。我们生活在网络之中,被网络的形态所消费。看看全球的政治、经济、后殖民主义时代所形成的全球化的文化,这是不言而喻的。作品用来表达一种思考,提供一种我们认识现实世界的参照和手段。人类试图接近欲望,但欲望本身就是一个陷阱,就犹如我们追求地平线一般。

Internet is intrinsic in the logic of globalization, and plays an important role in the current tides of globalization and a core method for spreading information on globalization. The lifestyles and thinking structure in the past are gradually changing, and society is being smelted and decomposed with the new social modes coming into being. "Positive and Negative Internet" copes with such structural changes, and involves itself in it in an artistic way as well as expresses viewpoints toward its problems. An idea of "taking cyberspace awareness as a new ideology" has been gradually recognized by people, and such new ideology is characteristic of cybersm, and has an air of some trans-cultural forms, which tries to integrate national and state ideology, and the special properties of culture.

As for the forms, the works is mostly radiating. And the forms are connected with the knots on the two radiating ends. Such knots classify the forms and are open to the space, while the knots extend to the space from infinite forms, while each knot is individual and interdependent at the same time. It refers to some certain network system, and forms a huge and inclusive system. We are living in the network, and are consumed by the network forms. It is self-evident for us in case we just have a look at the global cultures formed in the global politics, economy and post-colonial times. The works is used to express a kind of thinking, and provides a reference and measure for use to learn about the real world. Man tries to approaches desire, while desire is a trap in itself, just like the horizon we pursue.

正负互联网,2012
宋刚(钢)(意大利)
综合材料
© 宋刚(钢)

Positive and Negative Internet, 2012
Song Gang (Italy)
Comprehensive Materials
© Song Gang

作品以太阳七色象征人类智慧的光芒，随机自由的线形组合和直角方硬转折的空间构建，呈现出人类感性和理性的活动，并以不断创新的追求，持续创造精神与物质的成果。

艺术与科学的互动与互补，是人类智慧与创造的载体，承载着人类迈向新的彼岸。

The works uses the seven colors of sunlight to symbolize the light of human wisdom, and the random and free combination of linear patterns and the spatial construction of right angular turns shows human sensory and rational activities, the constantly innovated pursuits, the spirit of constant innovation, and the material achievements.

The interaction and supplement of art and science is the carrier of human wisdom and creation, and carries man to a new bank.

七色方舟，2010
赵萌（中国）
不锈钢、喷漆
© 赵萌

7-Colored Ark, 2010
Zhao Meng (China)
Stainless Steel, Spraying Paint
© Zhao Meng

这是一幅手工上色的显微照片,画面上是澳大利亚原生短梗刺果雏菊的果实。它那竖立的刺毛和毛茸茸的果梗有助于风力播种。这些果子成就了艺术家的美意。

Hand coloured micrograph of a fruit from a short-rayed burr daisy native to Australia. The bristly hairs and feathery pappus rays may primarily facilitate wind dispersal. 2.8mm long. From fruit, edible, inedible, incredible, courtesy the artist, Wolfgang Stuppy & Papadakis Publisher.

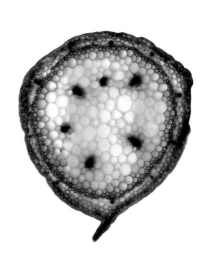
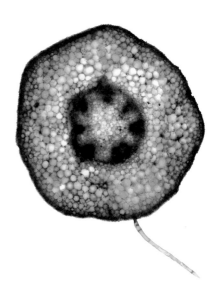

纸花葱,2011
罗伯·克塞勒
数码显微照片
© 罗伯·克塞勒

Allium Neapolitanum, 2011
Rob Kesseler
Digital Print on Paper
© Rob Kesseler

清晨 5 点，在北纬 79°（北极地区）的严寒中，周围还是一片漆黑。作者们把投影机安装在船舷上缘，与一座跟他们的船桅一样高的冰墙相距 20 英尺。就在破晓的一刻，投影与冰川正面日光的强度刚好匹配。

这组"冰雪文字"全部在北极高纬地区制作而成，总共 25 套，每一套都表明了北极正在消亡，而我们人类狂热的活动正是使这些美丽融化的起因。这些文字体现了我们将会失去什么，也表现了我们将传给后代一个什么样的世界。气候便是文化，而艺术则是积极的干预。

5a.m. and -10°C on the 79th parallel north, still dark. The authors had their projector mounted on the gunnels, 20 foot from a towering wall of ice, as high as the ship's mast. At the cusp of dawn, the projection matched the intensity of the daylight on the glacier front.

The series of "Ice Texts", all made in the High Arctic, are 25 in number and each referring to the disappearing arctic and how feverish human activity is causing this beauty to melt. The texts refer to what we stand to loose and the legacy we are handing to our children. Climate is culture and art is an active intervention.

冰雪文字——消融未来，2009
戴维·巴克兰、艾米·博尔坎
数码照片印刷品
© 戴维·巴克兰、艾米·博尔坎

Ice Texts: Discounting the Future, 2009
David Buckland, Amy Balkan
Digital Photographic Print
© David Buckland, Amy Balkan

这件作品是一次面部重建手术研究项目的产物，旨在确立"正常的"面部外貌参数，而得出的结论则是正常的外貌等于正常的功能。

The product of a research project conducted with a facial reconstructive surgeon that aimed to establish the perameters of " normal " facial appearance , the project concluded that normal appearance was normal function .

面部景观, 2012
斯蒂芬·法辛
数码印刷
© 斯蒂芬·法辛

Facial Landscape, 2012
Stephen Farthing
Digital Print
© Stephen Farthing

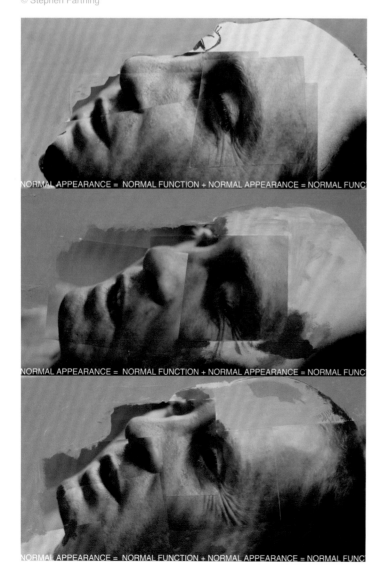

《家族谱——镜框》探讨天生与养育以及血统的观念。作者想要制作一套空白的家族谱，参观者可以在其中想像自己亲属的形象。意图是要创作一种在乐观与悲观之间盘旋的画面：这些镜框是等着装入亲属的照片呢，还是里面的照片已经被取掉了？

"Family Tree—Frames" relates to ideas of nature versus nurture and of genealogy. The author wanted to provide a blank family tree onto which the viewer could project his or her own relatives. The intention was to make an image that hovers between optimism and pessimism; are these frames waiting to be filled or have been emptied?

家族谱——镜框，2009
保罗·科德维尔
屏幕印刷
© 保罗·科德维尔

Family Tree—Frames, 2009
Paul Coldwell
Screen Print
© Paul Coldwell

《植物科学》这件作品探究人类与植物世界的关系，尤其是植物与医药的关系。这三张照片是在伦敦南部地区一个废弃的生物实验室中拍下的。

"Plant Science" is a performance project exploring mankind's relationship with the botanical world and in particular the relationship between plants and medicine. The three photographs were taken during a "visual rehearsal" in set of abandoned laboratories in South London as part of an ongoing residency that will lead to a series of live events in London in 2013.

植物科学，2012
佛斯特、海伊斯
数码照片印刷品
© 佛斯特、海伊斯

Plant Science, 2012
Forster, Heighes
Digital Photographic Print
© Forster, Heighes

作品于2012年在伦敦大学国王学院废弃的植物科学实验室拍成。这里曾因众多的科学发现而名满天下，云集了全球首屈一指的科学家们；如今这片废弃之地只剩它的精神还依然留存。这组空间作品尝试为实验室注入新的生命与温暖，希望历史不会完全湮没无闻。

These photographic installations were made in 2012 in the abandoned Plant Science Laboratory of Kings College, London. It was a site once famous for its scientific discovery and the home to many world leading scientists. Now in its abandoned state what is left is only its spirit. The installations in the space are an attempt to breath new life and warmth into the laboratories in the hope that the history is not entirely forgotten.

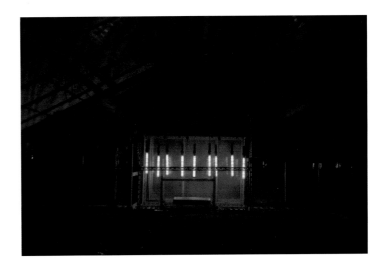

植物科学实验室，2012
克里斯·韦恩莱特
数码照片
© 克里斯·韦恩莱特，伦敦艺术大学

Plant Science Laboratory, 2012
Chris Wainwright
Digital Colour Photographs
© Chris Wainwright, University of the Arts London

《时间与潮汐》是在一艘船的货舱中完成的一系列组画,是与大河之间流动的关系的余痕,尤其是船舶往来的浪迹、风的魄力以及潮汐的起落。创作大多是在重大的日子和时间里完成的,最近一次是在 2012 年 3 月 6 日庆祝女王登基 60 周年的游船盛会期间。

"Time and Tide" are a series of drawings made in the "hold" of a ship. They are the residual traces of a dynamic relationship with the river, in particular the wash from passing vessels, the force of the winds, and the flow of the tides. The drawings are often made on significant dates, and times, the most recent on 03.06.12 during the River Pageant to celebrate the Queen's Jubilee.

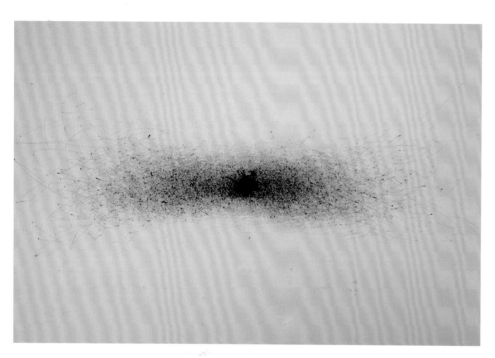

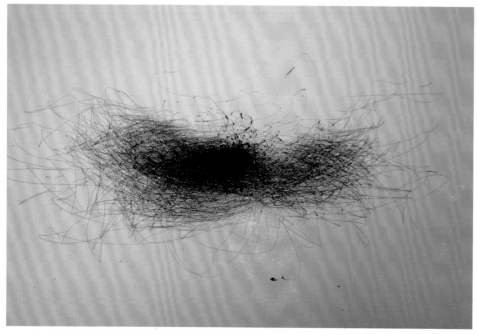

时间与潮汐,2012
安妮·丽迪亚特
纸质油墨画
© 安妮·丽迪亚特

Time and Tide, 2012
Anne Lydiat
Ink on Paper
© Anne Lydiat

在这件作品中，窗户是装置的焦点，把人们的注意力引向特定的主题，通常都有一帧照片或是一系列带着隐喻意味的物件放在窗户中或是悬挂的窗框上。这幅作品将艺术家们在秘鲁亚马逊探险时的照片放大并分割开来。其中一扇窗门和百叶都打开的西班牙窗框安置在"爱丽丝漫游奇境"式的旅途中，展现了亚马逊的神奇风貌。

In the works of Lucy + Jorge Orta, windows are focal points drawing our attention to a specific subject or issue. They usually find a photographic reference and a series of metaphorical objects inside or suspended from the structures. In this works the artists' photographs from their expedition to the Peruvian Amazon are enlarged and fragmented. The Spanish window frame, whose doors and shutters open in an "Alice in Wonderland" voyage, reveal the different Amazon landscapes.

世界之窗——亚马逊, 2010
露西·奥尔塔、豪尔赫·奥尔塔
木制窗框（带百叶）、Lambda 照片、层压德邦板、瓶子
© 圣吉米尼亚诺常青画廊 / 北京 / Le Moulin

Window on the World—Amazonia, 2010
Lucy Orta, Jorge Orta
Wooden Window Frame with shutters, Lambda Photographs, laminated Dibond, Plasma bottles
© Courtesy of the Artists and Galleria Continua San Gimignano / Beijing / Le Moulin

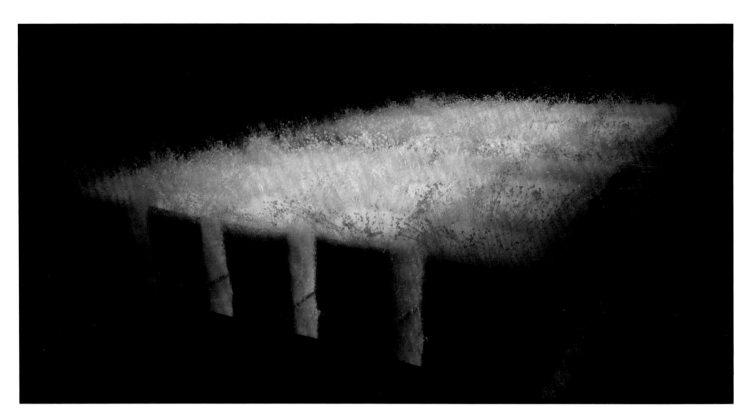

凝固的绽放,2010
岳嵩(中国)
化纤
© 岳嵩

Solidified Blooming, 2010
Yue Song (China)
Chemical Fiber
© Yue Song

《凝固的绽放》是一组用服装辅料子母扣做成的作品。这种材料拥有一种动感,它们自身的弹性象征着一种生命力,这种生命力似乎在流动与静止间徘徊。子母扣的两端仿佛是生活中的两种极端状态,也是凝固与绽放的同在。当将很多根透明的子母扣堆放在一起,呈现出一种"怒放"的美。在阳光之下,一根根细线连接着一个个触点,折射着太阳的点点光辉,轻轻触碰,它们便微微颤动,一股生命的力量由内而外缓缓倾泻。将这种美好寄托在材料之中,通过凝固与绽放的对比与碰撞,让自己的感情有了永恒的寄托,也让一切"凝固"在"绽放"时的灵动瞬间,记录下瞬间绽放的感动。

"Solidified Blooming" is a group of works made of snap button, an auxiliary clothes material. Such material is dynamic in itself, and their own elasticity symbolizes a vitality, which seems to wander between flowing and quietness. The two ends of the snap button seems to be the two extreme statuses of life, that is solidification and blooming. A kind of "bursting" beauty is displayed when many transparent snap buttons are piled together. The tiny lines are connected with the touch point, and refract the splendors of the sun under the sunshine, which will vibrate gently with a light touch, and a power of life will pour slowly from inside to outside. One's own feelings can have a permanent sustenance through placing such beauty on the materials, and comparing and colliding of solidification and blooming. And all will be "solidified" at the instant of "blooming" to record the instantly blooming impression.

童男、童女象征中国神话传说，传说本身亦是文化的传奇。现代科技又使童男、童女微缩变身，再造传奇。

文化传奇与科技传奇，共创人类文明的传奇。

Virgin boy and virgin girl symbolize Chinese myths and legends, while legends are the legend of culture themselves. Modern technology has displayed microform and transformation of the virgin boy and virgin girl to create legends again.

The cultural legend and technological legend jointly create legends of human civilization.

传奇，2011
周尚仪（中国）
青铜、银
© 周尚仪

Legend, 2011
Zhou Shangyi (China)
Bronze, Silver
© Zhou Shangyi

作品吸收了中国水墨的黑白韵味和意境，蕴有老子的"有无相生、黑白相倚、阴阳流变"的哲学意味，追求单纯中的丰富、虚空中的气韵。抒发了洒脱、高贵和融于天地之间的情怀。

作品采用了大写意的手法，以中国丝、水纱为材料。水纱的特性是轻、薄、透，材料本身已经传递出一种自然纯净之美，而且也与传统文人画的清、远、静相暗合。在材料制作、处理方面，利用手工纺织的丝质水纱透感特性和可塑性强的特性，进行不同层次的手工制作及叠加变形，使本来相对单一的轻薄柔软的水纱产生了内在的张力，增加了材料的丰富性。创作过程具有很大的随机性，追求一种随机自由的美。

The works has absorbed the black-white charms and mood of Chinese water and ink art, containing the philosophic senses of Laozi, i. e. "Extremes meet, black and white depends on each other, while Yin and Yang alternate". And it pursues richness in simplicity, and charms in emptiness, and expresses the unrestrained, noble feelings of being integrated into the world.

The works adopts freehand brushwork, and takes Chinese silk and gauze as the materials. Gauze is characteristic of light, thin and transparent, and conveys a natural and pure beauty, and coincides with the pureness, remoteness and quietness of conventional literati paintings. As for materials manufacturing and disposal, it carries out handwork and overlapped morphing at various levels by making use of the transparency and strong plasticity of silk gauze, thus producing intrinsic strain in the relatively monotonous light, thin and gentle gauze, and improving the richness of the materials. The course of creation involves great randomness, and seeks a random and free beauty.

清、远、静，2012
李薇 / 山东如意科技集团有限公司（中国）
真丝绡、水纱、光导纤维、其他材料
© 李薇

Pureness, Remoteness, and Quietness, 2012
Li Wei/Shandong Ruyi Science & Technology Group Co., Ltd. (China)
Silk Chiffon, Gauze, Photoconductive Fiber, Other Materials
© Li Wei

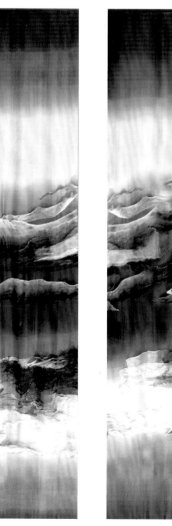

作品设计取材于内蒙古传统的佩饰，反映出草原游牧文化的浓郁的装饰情调。材料结合山东如意科技集团有限公司精选的"Baby Cashmere"，运用获得国际科技进步一等奖桂冠的世界独一无二的"如意纺"技术，将传统文化与科技文化、传统文化与时尚文化完美结合。

The design of the works draws on traditional Mongolian ornaments, and refllects the rick decorative charms of grassland nomadic culture. The choice "Baby Cashmere" of Shandong Ruyi Science & Technology Group (China) are jointly used with the unique "Ruyi weaving" technology that has won international first prize for technological progress for perfect combination of conventional culture and technological culture, and conventional culture and fashionable culture.

羊绒围巾设计——如意长歌，2012
张宝华 / 山东如意科技集团有限公司（中国）
羊绒
© 张宝华

Design of Cashmere Scarf—Intone of Ruyi, 2012
Zhang Baohua/Shandong Ruyi Science & Technology Group (China)
Cashmere
© Zhang Baohua

此情可待成追忆, 2009
杜大恺（中国）
水墨纸本
© 杜大恺

Feeling for Memory to Recall, 2009
Du Dakai (China)
Water-Ink on Paper
© Du Dakai

无论古今，自然的影响都是深刻的，古代尤甚，以至完全受自然支配，选择生产与生活方式。

江浙一带枕水而居，以木筑屋，以稻为食，以舟代步，因其多水少山，河网密集之故也，因势利导，久而久之，渐成生息相成的习惯与风俗，故有水乡，常言说"一方水土养育一方人"，诚亦哉焉。

生产力的发展不断改变人与自然的关系，物侯之异的支配性亦在发生变化，依水乡言之，筑屋行步已不再唯木、舟是从，习惯正趋淡化，风俗渐成记忆，面对于此，人们亦喜亦忧，得失之间，破费周折，似难两全。《此情可待成追忆》乃记其今日之所见所思矣。

The influences of nature are always profound in the past or nowadays, especially in ancient times, and even the means of production and living were chosen under the complete control by nature.

People in Jiangsu and Zhejiang Province live by the waters and build houses with wood feed on rice, and travel by boat, for there are many waters but few mountains, and a concentrated river network. Thus the habits and customs have gradually come into being as time goes by due to the situation, and it is often said in the water towns that "People are supported by the land where they live".

The development of manufacturing force constantly changes the relation between man and nature, and the dominance caused by the differences of substances is also changing. As far as the water town is concerned, people there have not only chosen wood and boat for house building or travelling, for such habits are fading out, and customs are gradually fall into memory. People feel both happy and worried at the sight of that, and it seems painstaking to have regard for both parties in face of gain and loss. "Feelings for Memory to Recall" records what one sees and thinks about today.

自从有了数码相机后,多少年来靠速写记录生活的习惯似乎在瞬间被改变了。

快速、简便地记录随时可见的图像,成为人生中的一大乐趣。于是,捕捉大地、天空,乃至时光的瞬间变化的欲望无处不在、无时不有。

《瞬间》即源于瞬间。

The long-time habit of recording life with shorthand has been instantly changed by the occurrence of digital camera.
Recording the visible images rapidly and easily from time to time has become great pleasure in life. Thus the desire to capture ground, sky and the instant changes of time has been omnipresent at any time.
"Instant" originates from instants.

瞬间, 2011
林乐成(中国)
天然纤维
© 林乐成

Instant, 2011
Lin Lecheng (China)
Natural Fiber
© Lin Lecheng

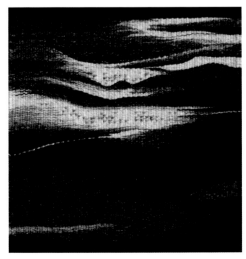 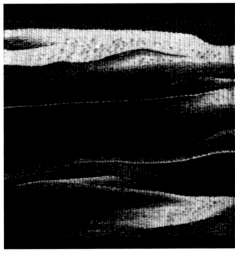 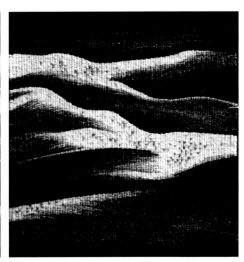

《光的蒙太奇》系列面料作品运用独创的复合材料和激光切割为手段，创造出可随时改变面料肌理和图案的新型科技面料。它通过改变反光材料的形状，收集光线，并改变光的方向，创建出一个以色彩、光线、影像三者结合的图案效果。让人们感受到一种新的图案视觉感受。

The series face material works named "Montage of Light" applies unique composite and laser cutting to create new technological face materials that can changes the texture and patterns of face materials at any time. It collects light and changes the direction of light through changing the forms of the reflecting materials, and creates pattern effect combining color, light and images, thus leaving new visual feelings toward patterns to people.

光的蒙太奇, 2011
黄思文（中国）
复合科技面料
© 黄思文

Montage of Light, 2011
Huang Siwen (China)
Composite Technological Face Materials
© Huang Siwen

《肯内梅尔沙丘公园》是两位艺术家在北荷兰肯内梅尔沙丘国家公园的七个地点拍摄的纪录片，时间跨度刚好一年，从某年1月到次年1月。艺术家们每周在从同一个位置反复拍摄每个场景，拍摄时间都是在中午前后。在使用定制的软件后，每一组镜头都被编辑为流畅的过渡画面，可视地表现了季节与景物的变化，使人们能够从感官的层级上体察日常生活中绝不可能直接察知的变迁过程。艺术家们在摄影现场体验到的宁静还依然驻留在拍成的影片上。

"Kennemerduinen" is a project for which the artists documented seven locations in the National Park, the Kennemer Dunes, in the province North-Holland. Each film has a duration of nine minites and covers exactly one year, from one January to the next. On a weekly basis, every scene was repeatedly photographed from the same position and at the same time of day, around noon. With custom developed software, each series of shots was edited into fluid transitions to make changes in seasons and landscape visible. Slow processes of transformation that are never directly perceptible in daily life, are made perceptible on a sensory level. The calmness—that one experiences on the site concerned—is still present in the resulting films.

肯内梅尔沙丘公园，场景 H, 2011
德里森、维尔斯塔彭（荷兰）
数字影片
© 德里森、维尔斯塔彭

Kennemerduinen, Scene H, 2011
Driessens, Verstappen (Netherlands)
Digital film, Mpeg 4 file on USB-stick
© Driessens, Verstappen

熏风撩动人心，霞光洒落，平静的湖水波光粼粼，这些是雕塑创作灵感。
雕塑有无数个不锈钢碎片组成，象征着一种团结、凝聚。碎片灵活可动，随风吹过，会发悦耳的声音。
雕塑关注生态，用一种工业化的方式表现自然之美，提醒人们在忙碌的工作之余反省自我，体味自然与生命。

The quiet lake ripples shines under the inviting warm breezes and sprinkling glow of the sun, and such are the inspirations for creating sculpture.
The sculpture is made up of countless fragments of stainless steel, which symbolize union and cohesion. The fragments are flexible and mobile, and will send out pleasant sounds with the passage of wind.
The sculpture cares about ecology, and expresses natural beauty in an industrial way to remind people to reflect on themselves after the busy work, and experience nature and life.

光音，2012
李鹤（中国）
不锈钢
© 李鹤

Light and Sound, 2012
Li He (China)
Stainless Steel
© Li He

在这个快节奏的时代，物质成为一部分人追求的目标，成为衡量人生价值的重要体现，尤其对奢侈品的一种顶礼膜拜。这件以人工材料玻璃和不锈钢制作的巨大戒指正表现了人们日益膨胀的物欲。人类产生越多的消费，自然就越匮乏。

Materials have become the object to seek, and an important embodiment to measure life value in the times of rapid rhythm, especially, the worship of luxurious goods. The huge ring made of artificial glass and stainless steel shows the increasingly expanding human desire. The more consumption man produces, the more deficient the natural will be.

物语，2009
刘立宇（中国）
玻璃、不锈钢
© 刘立宇

Monogatari, 2009
Liu Liyu (China)
Glass, Stainless Steel
© Liu Liyu

上小学的时候，就喜欢红颜色。图画课上，用过红药水画红旗飘舞，因为那红药水既透明又有黄色的力量隐含在里面。大学本科的时候，被漆画工作室朱漆的颜色所吸引，毫不犹豫地选择了漆艺术专业。至今，每每看到朱漆，仍然令作者心驰神往。

朱漆，红火、赤烈。

金箔，发光，高贵。

新生的活力，太阳般的炽热。

还是赤色、金色。像蜗牛，又像小鸟……寻寻觅觅。以赤者满腔热情之诚，以金者千锤百炼之信，寻觅每一个生命过程的真谛和永恒。

The author liked red color when he was studying in primary school. He once painted flying red flags with mercurochrome in the drawing class, for the mercurochrome contains transparent and yellow power inside. Later, he was attracted by the color of the red paint in the lacquer painting in his university, so he chose lacquer art major without any hesitation. And he still have a deep longing for red paint at the sight of it today.

Red paint is red and fierce.

Gold leaf, brilliant and noble.

Newly-born vitality, and burning like sun.

Still red and golden. Search something like snail and bird…; Seek the truth and permanence of each life with the warm sincerity of red and tampered confidence of the gold.

《立》、《觅》，2010
周剑石（中国）
大漆
© 周剑石

"Standing", "Search", 2010
Zhou Jianshi (China)
Chinese Lacquer
© Zhou Jianshi

科学家讲时间是永恒的，
也有科学家讲时间会终结的。
但至少不是现在，
今天的时间依然在运行，
明天的时间依然会运行。
我们既看不到时间的终点，
也看不到它的终结，
我们只看到了星移斗转。

Some scientists say time is eternal, while some say time will end. Sometime; but at least not now, for time still goes on today and will still do so tomorrow. We cannot see the ending point of time or its end. We have only seen the passage of time.

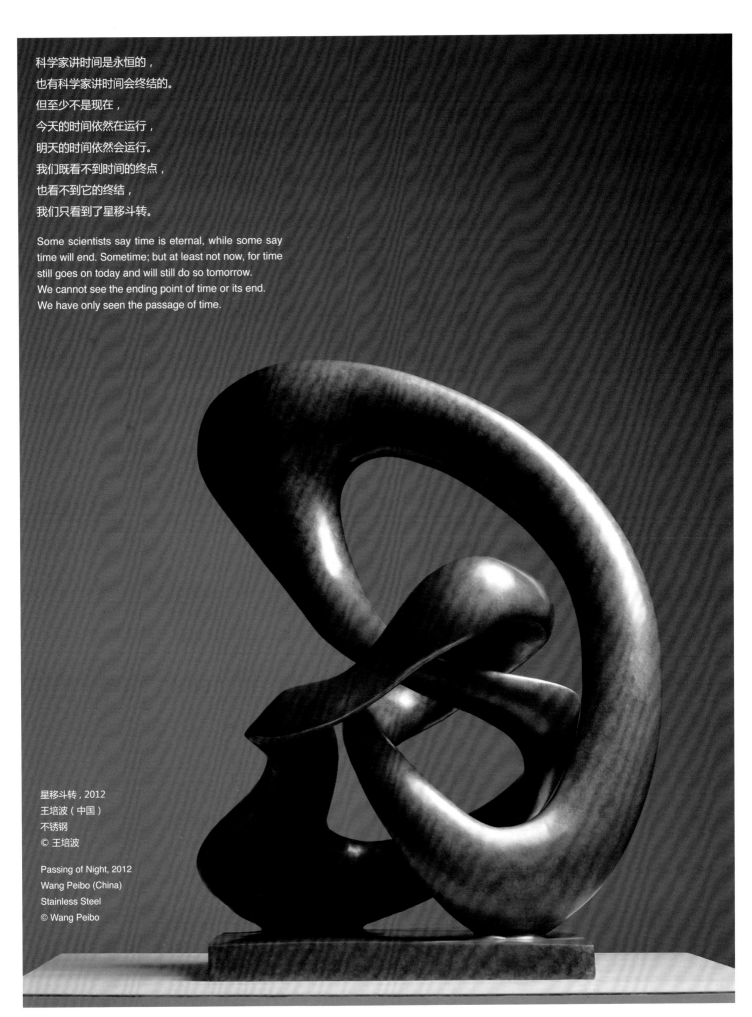

星移斗转, 2012
王培波（中国）
不锈钢
© 王培波

Passing of Night, 2012
Wang Peibo (China)
Stainless Steel
© Wang Peibo

我们所熟悉的日常物件已经被两种化学加工程序所改变：蚀刻（金属溶解）与电镀（金属沉积）。假如我们先来处理十把黄铜钥匙铸件，在蚀刻加工过程中，五把钥匙被浸没在一种蚀刻液中，其中一把钥匙时不时地从溶液中移开，直到它们最后形成一个小小的、面目全非的形状。而在电镀加工过程中，我们通过电解作用将硫酸铜溶液的金属原子沉积到五把钥匙上。与蚀刻一样，五把钥匙中有四把所经过的化学处理时间更长一些。另一把钥匙不经过电镀处理，保留原状。蚀刻与电镀通常在工业中用于表面处理，但是在上述这种处理过度的情况下，结果会变得难以预料、无法控制。我们以连续的序列来呈现这两种工艺过程逐渐加强的效果，它凭空产生了一种常见的形状，然后又让这件古怪的赘生物最终消失于无形。

The recognisable form of every-day objects has been changed through the effects of two chemical processes: etching (dissolving of metal) and galvanising (deposition of metal). First, ten casts are made in brass of a key for instance. During the etching process five keys are submerged in an etching solution. From time to time one of the objects is removed from the solution, until finally a small unrecognisable shape was left over. During the galvanisation process, metal atoms from a copper sulphate solution are deposited onto the objects by means of electrolysis. Just like during the etching process, four keys have been subject to the chemical process for an increasing period of time. One key is not treated but kept in its original state. Usually, etching and galvanising are applied as an industrial surface treatment but whit the extreme exposures that are applied in this case, the results have become unpredicable and uncontrolable. The gradually increasing effect of the two opposite processes has become visible in a continnual form sequence, in which the recognisable shape emerges from nothing, to finally disappear again under a whimsical excrescence.

变形 13 号——花生, 2003
德里森、维尔斯塔彭(荷兰)
10 件镀镍铜铸件
© 德里森、维尔斯塔彭

Morphotheque No.13, Peanut, 2003
Driessens, Verstappen (Netherlands)
Nickel-plated brass, 10 elements
© D&V

作品选用天然羊毛毛毡，以普洱茶为植物染色剂，制作八幅系列壁挂，表现自然界的优美景象。

"赋格"来源于拉丁语，原意是追逐和飞翔，是一种复音音乐的创作形式。

"风的赋格"以毛毡的厚重温暖、普洱茶的色与味、国画的卷轴形成主题片段，借用主调和属调的有机重复，唤起两位作者在同一主题下的应和奏答，引发艺术情感的共鸣，于对话和交流中呈现和而不同的艺术表达。作品调动多种感官生活记忆，以自由、快乐、易参与的创作形式提供艺术介入生活的多种可能，纾解现代社会对心灵的压力，倡导人们回归自然。

Natural wool felt is adopted for the works, and Pu'er tea is taken as its plant coloring agent to make eight series hangings and present the beautiful natural sceneries.

"Fuga" comes from Latin, which originally means pursuit and flight, and is a means of creating composite tones. "Fuga of Wind" takes the heaviness and warmth of the wool felt, color and tastes of Pu'er tea, and the reel of traditional Chinese painting to form the thematic sections, and draw on the organic repetition of the homophony and the dependent tones to awaken the echoes and responses in performance by the two authors under the same theme, and arouse the consonance of artistic feelings as well as demonstrate harmonious but various artistic expressions during talks and exchanges.

The works activates multiple sensory memories of life, and provide various possibilities for involvement of arts in life with free, joyful and easily accessible means of creation to ease up mental pressure from modern society and call for people to return to nature.

风的赋格，2012
吴波、朱小珊（中国）
毛毡、丝绸、普洱茶、鸡翅木
© 吴波、朱小珊

Fuga of Wind, 2012
Wu Bo, Zhu Xiaoshan (China)
Felt, Silk, Pu'er Tea, Chicken-Wing Wood
© Wu Bo, Zhu Xiaoshan

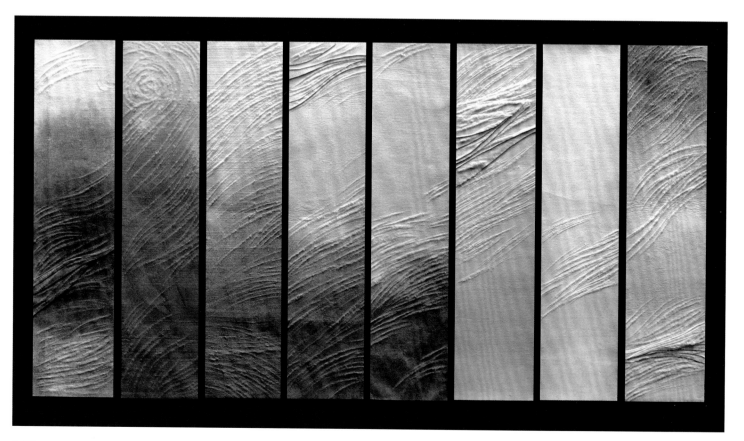

从最朴实的造物情感出发，借助服装的语言，发掘传统制毡工艺的智慧能量。以最传统的手工技艺表达当今的时尚态度。

The works taps out the intelligent energy of conventional felt manufacturing technology from the simplest feelings toward manufacturing by means of taking clothes as the language, and express the contemporary fashions and attitudes with the most conventional craftsmanship.

毡衣无缝，2012
李迎军（中国）
羊毛
© 李迎军

Seamless Felt Clothes, 2012
Li Yingjun (China)
Wool
© Li Yingjun

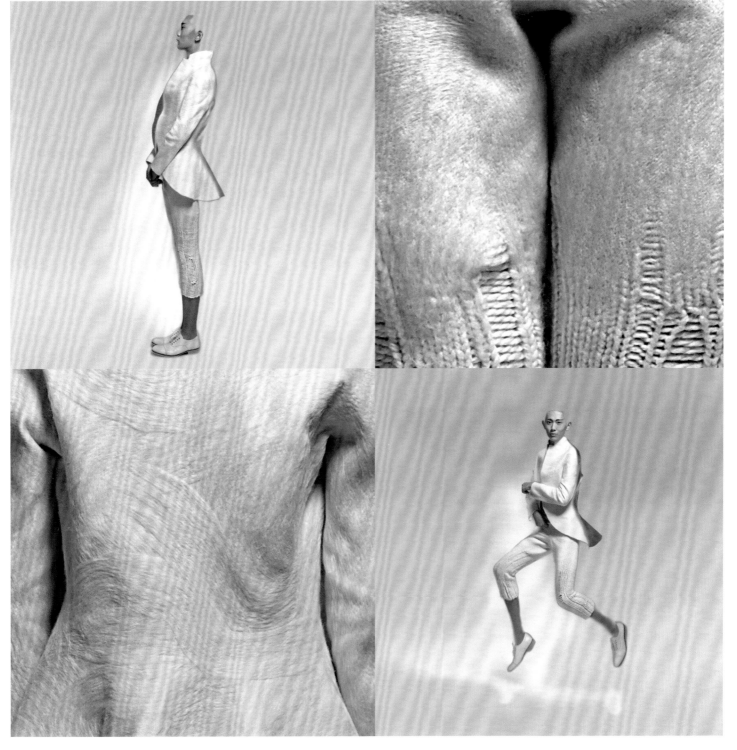

作品《恍若隔世》上、下两幅表现了战争、海啸、地震灾难等对地球和人类造成的生态危机，警示我们高科技的发展在给社会带来进步文明的同时，是不能以对地球的破坏和虐夺作为代价的，否则，未来的地球将不会有生态的家园。中间一幅把银河系中安宁的月球作为人类的理想幻境，寄托着一种对精神家园的守望。

The two sections of the works "Seeming Lapse of Whole Generation" shows the ecological crisis caused by war, seismic sea waves and earthquake etc. on the earth and man, and warns that we shall not seek social progress and civilization with the high-tech development at the cost of destruction and abuse of the earth, otherwise there will be no ecological family on the earth in the future. The picture in the middle takes the quiet moon in the galaxy as the ideal wonderland of man, and rests expectation for spiritual family on that.

恍若隔世，2012
陈辉（中国）
水墨纸本
© 陈辉

Seeming Lapse of Whole Generation, 2012
Chen Hui (China)
Water Ink on Paper
© Chen Hui

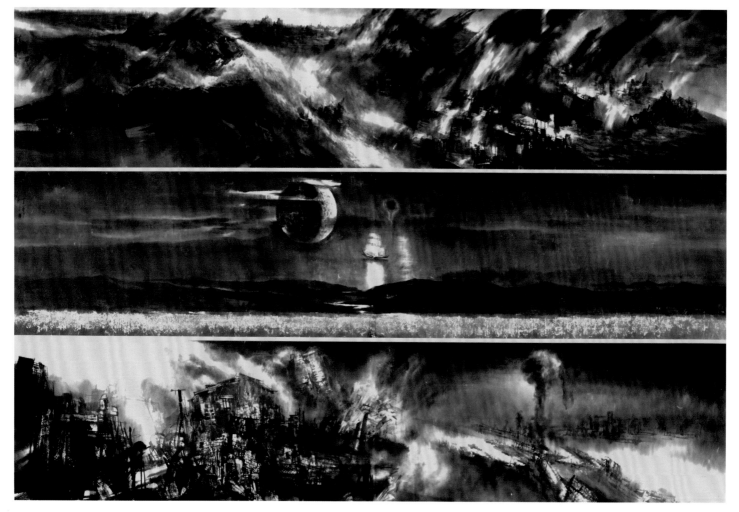

草、野草，红橙黄绿黑白灰的大地的光谱，居无阔，心无界，微妙而苍茫，大有万仞之势，通向天宇，小而入乎精微，它像所有生命一样，精细博大，在现代几万倍的显微镜下，草叶的结构一层一层严密有序，充满神秘。在上古神话的世界里，它又是夸父追日逮之禺谷、道渴而亡后，其毛发化为野草的神奇，长满大地。故草心怀日月，生生不息，其内心充满了艺术与科学的人文精神。

Grass and weeds, the optical spectrum of ground of various colors, are vast, boundless and subtle, which may be huge and powerful, and lead to the sky, or tiny and refined. It is tiny and vast like all the other lives, and the structure of the grass leaves are of strict and orderly layers, and is full of mysteries. As noted in the ancient myths, it is marvelous in that the hair of Kuafu changed into omnipresent weeds after he died of thirst on its way to Yugu Valley. Thus the grass boasts the vast universe and eternal growth, and is full of humanistic spirit on art and science inside.

追日草，2012
刘巨德（中国）
水墨纸本
© 刘巨德

Sun Pursuing Grass, 2012
Liu Jude (China)
Water Ink on Paper
© Liu Jude

《表面》是由九组不同色彩、不同图纹组合、垂直立面组成的作品。它们就像舞台的翅膀,在展翅翱翔。从早晨,到下午,再到晚上,不断变换的光线,映照在每一个立面的装饰层上,展现出不同的光影效果。这个由地板构成的巨大"雕塑",凸显出一种记忆与未来的对比,代表了地面产品发展的"未来主义"。

The line of "Surface" consists of 9 works displaying "Surface" through different colors, pattern combinations and vertical facades. Like wings of the stage, they make it seem to be soaring. The changing lights from the morning to the afternoon and till the evening shine upon the decoration layer of each façade and create different lighting effects. This huge "sculpture" made up of the floor highlights a comparison between memory and future, representing futurism of floor product development.

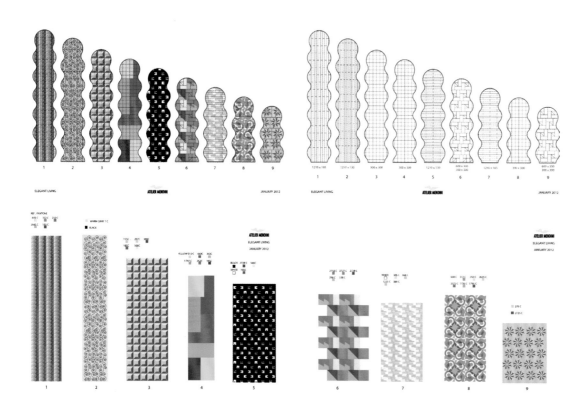

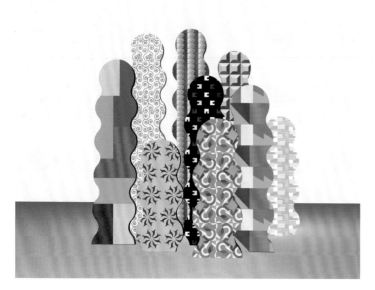

表面,2012
阿历山德罗·门迪尼(意大利)
木制品
© 阿历山德罗·门迪尼

Surface, 2012
Alessandro Mendini (Italy)
Woodware
© Alessandro Mendini

《明天会更好》是作者花了多年时间制作的一套切割纸组画，它们描绘了被自然灾害和人为灾难毁坏的家园，其中的住宅形象是作者过去收集起来作为绘画主题的。他有意使这套组画成为昙花一现的描摹。作者对作品中的住房甚感兴趣，就如同喜欢黑色电影中的人物一样。这组绘画展现了某个家园在被送到填埋场之前最后的时刻，它们这样过早地被埋葬，这是任何人始料不及的。通过在住宅之中日常的琐事，住宅与居住者之间便产生一种不断变化发展的关系。一场悲剧性事件会损害住宅的稳定性，从而造成居住者情感的动荡。这些住宅会像它的居住者那样有生有死，而不是屹立不倒的坚固建筑。住宅随着它的外部环境和它与居住者的对话而形成特殊的性格，从而通过记忆和互动与居住者结成纽带。我们对物理环境有什么样的依附感？这些感觉会让我们坚强或是使我们软弱？我们把房子当作私有财产，以为它们是理所当然的栖身之所，但如果从我们生活中抽走它坚固的特性，它就能揭示出我们的脆弱。这组画还进一步探讨了我们居家生活的问题，这也是作者一直在努力表达的内容。"建筑结构与家庭结构的关系"是作者在今后的作品中将进一步去探讨的主题。

"Tomorrow will Get Better" is a series of drawings with cut paper that the author has been working on for the past couple years. The images depict homes that have been demolished from various disasters both natural and man-made, that he has been collecting and using as drawing subjects. He intend for the drawings to become ephemeral portraits. He has been interested in houses in his work, as a character in a narrative, similar to a Gothic Noir film. The present series exhibits these final moments of a home before it is sent to a landfill, a premature burial that no one ever expects. A relationship between the house and inhabitants develops and evolves through daily events that take place within the confines of a home. A tragic event will jeopardize the physical stability of the home, resulting in the emotional instability of its inhabitants. The home has the potential to live and die just like its inhabitants instead of a solid structure that is built to survive through generations. A home develops character with its exterior surroundings as well as its interior dialogue with the characters that create a bond through memories and interactions. What are our attachments to physical surroundings? How do they strengthen or weaken us? Houses being taken for granted as private property with equity and security have the potential to reveal our vulnerability when the sound character is subtracted from our lives. The series further investigates issues that encompass domestic life that the author has been working toward. The relation of architecture to our domestic structure is a theme that he will continue to pursue in future work.

明天更美好，2007—2009
马修·科克斯（美国）
激光切割纸、迈拉聚酯薄膜、石墨
© 马修·科克斯

Tomorrow will Get Better, 2007-2009
Matthew Cox (USA)
Laser cut paper, Mylar, Graphite
© Matthew Cox

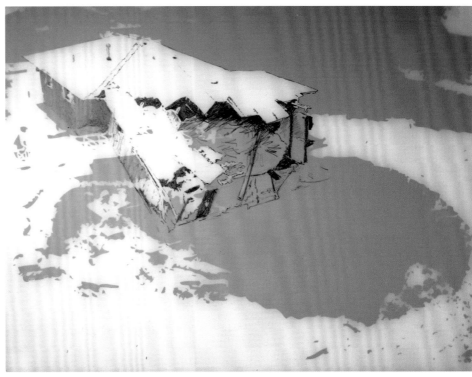

187

我们从哪里来？我们到哪里去？我们是什么？这些千百年来的问题，在21世纪的今天还会问下去。我们被工业文明和科技成果包围甚至异化。但无论人类文明如何演进，人类的本性和良知是不变的。那就是：爱情、正义、忠诚、宽容。

本作品借用符号、文字和图形来塑造当今人类形象，意在表现当今世界是由多元文化、多种社会和多种生活方式构成，试图传达丰富的时代信息并追求永恒的人性。

Where did we come from? Where will we go? What are we? The eternal questions will be further asked in the 21st century. We have been surrounded or even alienated by industrial civilization and technological achievements. Human nature and conscience remain changed no matter how human civilization evolves. That is love, justice, loyalty and tolerance. The works builds up current human images with symbols, characters and pictures, and is intended to show the current world is made up of pluralistic cultures, multiple societies and lifestyles and tries to convey rich information on times and seek eternal human nature.

我们，2008
潘毅群（中国）
不锈钢
© 潘毅群

We, 2008
Pan Yiqun (China)
Stainless Steel
© Pan Yiqun

城市的变迁，运用抽象的形式表现具象的现实，线和面及立体空间的运用及信息内容的传递使人们有想像的空间。

Changings of City uses abstract form of concrete reality, such as line, plane, three-dimensional space and the use of information… to transfer what people have imagined.

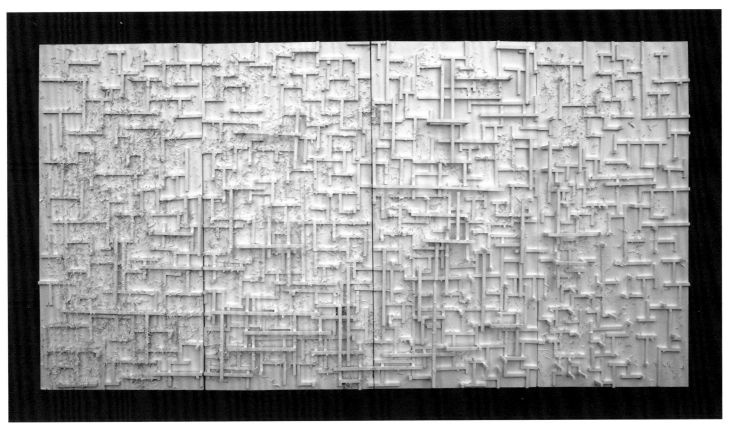

城市的变迁——构成2号，2012
洪兴宇（中国）
综合材料
© 洪兴宇

Changings of City, Form No. 2, 2012
Hong Xingyu (China)
Mixed Material
© Hong Xingyu

石画艺术是观照"天人合一"的哲学命题,也是体现"天之道"与"人之道"的和谐。天工是第一笔,人工是第二笔,二者合之即成石画。宇宙信息、天地生态、人文智慧都可以在一石之中展开。天人和谐之美是科学与艺术永远追寻的前沿;"天之道,利而不害;圣人之道,为而不争"是"尊道贵德"的总体路线。

The stone painting art reflects on the philosophical proposition of Harmony between man and nature, and also represents the harmony between the way of heaven and the way of human. The work of nature is the first stroke, the work of man is the second stroke, and combination of both makes the stone painting. The information of universe, ecology of man and nature, and humanistic wisdom may be developed in a stone. The beauty of harmony between man and nature is the frontier that art and science pursue at all times. The way of heaven is to benefit others and not to injure. The Way of the sage is to act but not compete. It is the general route of Respecting the Tao and Honoring Virtue.

天人合一,2012
杨中有(中国)
大理石板
© 杨中有

Harmony between Man and Nature, 2012
Yang Zhongyou (China)
Marble Slab
© Yang Zhongyou

后记

在 2001、2006 年两次"艺术与科学国际作品展暨学术研讨会"的积累与经验基础上,清华大学美术学院又全力组织策划了"第三届艺术与科学国际作品展暨学术研讨会"活动。这次活动以"信息·生态·智慧"为主题,是全球新文化语境下,艺术与科学的又一次盛会。

我们身处信息时代,以生态文明建设为目标的进程中,机遇与挑战并存。科学的突飞猛进,为人类带来了前所未有的可能性以及某些结果的未知性,重塑人文精神和可持续发展的理念已成为共识。艺术的审美和表现也越来越与科技相融合,科学与艺术日益成为时代的热点命题之一,日新月异的变化也正深刻影响和塑造着我们的生态。一方面体现在以创新为主导的科技文明、人文生态的蓬勃发展,另一方面又反映出某些自然生态的退化、困顿与挣扎。借此,以艺术与科学结合的角度和高度,面向人类未来,针对当下问题,凝聚思想智慧、激发创意、寻求解决方案,为实现人类文明与地球生态的和谐共生而努力,是我们这次活动的宗旨。

第三届艺术与科学国际作品展暨学术研讨会在这样的主题之下,通过公开展览及学术讨论的方式,总结艺术与科学间相互碰撞与融合的经验,力求揭示二者结合的巨大潜力;尝试不断拓展、深化两者间的研讨范畴,提高由此而激发的创新水平;也提出问题与思考,进而促进艺术与科学的和谐发展,并服务于整个社会文化。

本次活动筹备历时近两年。按照五年一届的传统,清华大学艺术与科学中心的努力和准备从未间断。时值 2011 年初启动展览的策划,恰逢清华大学百年校庆及首届北京国际设计三年展,后经学校商议决定,将艺术与科学展览推后一年,以便集中精力更加细致地筹备。一年后的 2012 年初春,第三届艺术与科学国际作品展筹备办公室正式成立。在清华大学美术学院各单位的密切配合下,一支由清华美院的青年教师和研究生组成的团队全身心地投入筹备工作中。展览筹备的艰辛,非身临其境不能真正体悟。

本次活动也得到了国外十多个国家，几十所高等院校、媒体艺术实验室、艺术与科学研究机构的积极参与，世界各地艺术与科学领域的最新成果云集北京。同时，这些新成果把中国对该命题的探索引入一个新的多元世界。这不仅为中国的艺术与科学事业进一步走向国际化、综合化提供了广阔的交流空间，也将为清华大学培养交叉领域高素质、高层次的创造性复合型人才奠定了坚实的基础。

策展及展览过程中，还有许多国家驻华使馆的文化处、教育处及各类基金会等在征集展品、邀请艺术家等方面，协助筹委会做了大量艰苦而细致的工作。山东如意科技集团、九牧集团、深圳市福田区政府、三林·生活家集团等还为本次展览活动提供了重要的资金支持；中国建筑工业出版社也为本次活动的作品集和论文集出版作出了贡献。

在此，我们向上述单位、机构与所有支持、帮助、指导本次活动的领导、前辈、专家、学者、老师及社会各界同仁致以诚挚的感谢。

清华大学美术学院院长

POSTSCRIPT

Carrying forward the accumulation of experience of the two terms of Art and Science Works International Exhibition (Symposium) in 2001 and 2006, the Academy of Art and Design of Tsinghua University now again make its full efforts to launch the 3rd Art and Science Works International Exhibition (Symposium). With "Information • Ecology • Wisdom" as its theme, this term of event will be a great event of art and science once again under the global new cultural context.

In the information era, opportunities and challenges coexist for the development of human civilization. The rapid progress of new media and science gives impetus to the integration of aesthetic expression and science, and gradually extends the antenna of thinking to new artistic propositions. The rapid changes happened day to day are also shaping our ecology, which is manifested, on the one hand, in the blooming of the humanistic ecology led by cultural innovation, and on the other hand, in the degradation, hardship and struggle of the natural ecology. Therefore, the combination of art and science embodies the wisdom of the ages. It serves our lives and will write a new chapter for the harmonious coexistence between human civilization and ecology.

Under such theme, the 3rd Art and Science Works International Exhibition (Symposium) is to summarize the experience of collision and integration between art and science through public exhibition and academic discussion, striving to reveal the great potential for combination of the two fields. It will try to constantly expand and deepen the scope of study in these two fields, and improve the level of innovation inspired by such expansion and deepening. In this way, it aims to further promote the harmonious development of art and science, and finally to serve the social culture as a whole.

The preparatory period of the event lasted nearly two years. According to the tradition of one term every five years, the Art and Science Research Center of Tsinghua University has never stopped their efforts and preparation. Initially, the planning of the exhibition was started in early 2011, which happened to coincide with the centennial celebration of Tsinghua University and the first term of the Beijing International Design Triennial. Then after consultation, the university decided to put off the Art and Science Works International Exhibition by a year, so that focused efforts will be made for detailed preparation. One year Later, in the early spring of 2012, the preparatory office of the 3rd Art and Science Works International Exhibition was officially established. Under the close collaboration among various units of the Academy of Art and Design of Tsinghua University, a team of young teachers and postgraduates of the academy dedicated themselves to the preparatory work, which was so arduous and laborious that one cannot truly understand unless personally on the scene.

This event has also been actively participated by dozens of institutions of higher learning, media arts labs and art and scientific research institutions from over ten foreign countries. The latest achievements in the art and science fields across the world will gather in Beijing. Meanwhile, these new achievements will lead China's exploration of propositions in these two fields to a new world of diversity. This will not only provide a broad space for exchanges for China's art and science causes to move further toward internationalization and integration, but will also lay a solid foundation for Tsinghua University to cultivate high-quality, high-level creative talents with comprehensive knowledge in interdisciplinary fields.

During the process of exhibition planning and exhibition, the cultural offices and education offices of many countries' embassies in China have also assisted the preparatory committee in such aspects as collecting exhibition works and inviting artists and have done a lot of arduous and meticulous work. The Shandong Ruyi Tech Group, JOMOO Group Co. Ltd., Government of the Futian District of Shenzhen, Samling Elegant Living Group, among others, have provided important funding support for this exhibition. China Architecture and Building Press has also made contributions to the publishing of the works collection and essays collection of this event.

Here, we would like to express our sincere thanks for the above-mentioned units and institutions and all leaders, seniors, experts, scholars, teachers and people from all walks of life who have provided their support, assistance and guidance for this event.

Lu Xiaobo
Dean of the Academy of Art and Design of Tsinghua University

图书在版编目（CIP）数据

第三届艺术与科学国际作品展作品集 / 第三届艺术与科学国际作品展暨学术研讨会筹备办公室编 . 一北京：中国建筑工业出版社，2012. 10
　　ISBN 978-7-112-14756-4

　　Ⅰ. ①第… Ⅱ. ①第… Ⅲ. ①建筑设计—作品集—世界—现代　Ⅳ. ① TU206

中国版本图书馆 CIP 数据核字（2012）第 239004 号

总策划：鲁晓波
策　划：杨冬江
统　筹：蔡　琴　王旭东
审　校：王旭东　蔡　琴　王晓昕　刘润福　姜　申　阎　旭
编　辑：阎　旭　李诗雯　刘雅羲　宋　扬
协　调：管沄嘉　孟凡顺　贺秦岭　王　鹏
书籍设计：陈　楠
版式设计制作：阎　旭　郭宏观
责任编辑：唐　旭　吴　绫
责任校对：党　蕾　关　健

第三届艺术与科学国际作品展作品集
A WORKS COLLECTION OF THE 3RD ART AND SCIENCE WORKS INTERNATIONAL EXHIBITION
第三届艺术与科学国际作品展暨学术研讨会筹备办公室　编
＊
中国建筑工业出版社出版、发行（北京西郊百万庄）
各地新华书店、建筑书店经销
北京方嘉彩色印刷有限责任公司印刷
＊
开本：965×1270 毫米　1/16　印张：12¼　字数：430 千字
2012 年 10 月第一版　2012 年 10 月第一次印刷
定价：168.00 元
ISBN 978-7-112-14756-4
　　　（22841）
版权所有　翻印必究
如有印装质量问题，可寄本社退换
（邮政编码 100037）